Fabrication and Characterization of TiO2 Nanofibers for Energy Applications

By

M.V.SOMESWARARAO

CONTENTS

Chapter No.	Title	Page No.
Chapter 1	Introduction	1-32
Chapter 2	Literature Review and formulation of the thesis	33-61
Chapter 3	Fabrication and Characterization Techniques of Nanofibers	62-100
Chapter 4	Electrospinning Process Parameters Dependent Investigation of TiO_2 Nanofibers	101-130
Chapter 5	Preparation and Investigation of TiO_2/ZnO Composite Nanofibers for Photocatalytic Applications	131-150
Chapter 6	Conclusions and Future Remarks	151-154
	List of Publications	155

TABLE OF CONTENTS

		Page No.
	Declaration	i
	Certificate	ii
	Dedication	iii
	Acknowledgements	iv
	Preface	vi
	Table of Contents	ix
	List of Figures	xii
	List of Tables	xiv
	List of Abbreviations	xv
1	**Chapter-1**	**1-32**
1.0	Introduction	1
1.1	Background of Electrospinning	2
1.2	Electrospinning Process	3
1.3	Essential Factors for Electrospinning	7
1.3.1	Viscosity	8
1.3.2	Conductivity	9
1.3.3	Homogeneity	10
1.4	Applications of Electrospun Nanofibers	11
1.4.1	Stimuli-Responsive Applications	11
1.4.2	Shape-Memory Applications	12
1.4.3	Self Cleaning Applications	13
1.4.3.1	Superhydrophilic Surface	13
1.4.3.2	Superhydrophobic Surface	14
1.4.4	Self-Healing Applications	15
1.4.5	Living Applications	16
1.4.6	Sensing Applications	17

1.4.7	Energy Applications	17	
1.4.8	Filtrationor Separation Applications	18	
1.4.9	Textile Applications	18	
	References	21	
2	**Chapter-2 Literature Review**	**33-61**	
2.1	Literature Review on Electrospinning Nanofibers	33	
2.2	Formulation of the Problem and Organization of Thesis	49	
	References	54	
3	**Chapter-3 Fabrication and Characterization Techniques of Nanofibers**	**62-100**	
3.1	Fabrication Techniques of Nanofibers	62	
3.1.1	Template Synthesis	62	
3.1.2	Drawing	63	
3.1.3	Phase Separation	66	
3.1.4	Self-Assembly	69	
3.1.5	Electrospinning method	70	
3.2	Characterization Techniques	74	
3.2.1	X-Ray Diffraction (XRD)	74	
3.2.2	Ultra-Violet Visible Spectroscopy (UV-vis)	77	
3.2.3	Fourier Transform Infrared Spectroscopy (FTIR)	80	
3.2.4	Raman Spectroscopy (RS)	84	
3.2.5	Thermogravimetry/Differential Thermal Analysis (TG/DTA)	86	
3.2.6	Field-Emission Scanning Electron Microscope (FESEM)	89	
3.2.7	Transmission Electron Microscope (TEM)	93	
3.2.8	Electron Dispersive Analysis of X-rays (EDAX)	95	
	References	97	

4	**Chapter-4. Electrospinning Process Parameters Dependent Investigation of TiO$_2$ Nanofibers**		101-130
4.1	Introduction		101
4.2	Experimental Details		108
4.3	Results and Discussion		111
4.3.1	XRD Analysis		111
4.3.2	Surface morphology and EDX		113
4.3.3	TG/DTA Analysis		114
4.3.4	FESEM Analysis		116
4.4	Conclusions		126
	References		127
5	**Chapter-5 Preparation and Investigation of TiO$_2$/ZnO Composite Nanofibers for Photocatalytic Applications**		131-150
5.1	Introduction		131
5.2	Experimental Details		136
5.3	Results and Discussion		138
5.3.1	XRD Analysis		138
5.3.2	FESEM Analysis		140
5.3.3	Surface morphology		141
5.3.4	TEM Analysis		142
5.4	Conclusions		146
	References		148
6	**Chapter-6. Conclusions and Future Remarks**		151-154
	List of Publications		155

LIST OF FIGURES

Figure No.	Name of the Figure	Page No.
1.1	The schematic of typical electrospinning setup	3
1.2	Formation of Taylor cone by varying the applied voltage	4
1.3	Illustration of electrospinning jet path with the applied voltage	6
1.4	Overview of synthesis and applications of nanofibers	20
3.1	An illustration of template synthesis of nanofibers	63
3.2	Step-by-step drawing process for the preparation of nanofibers	64
3.3	Step-by-step drawing process for the preparation of nanofibers	66
3.4	Phase-separation approach for the preparation of nanofibers	67
3.5	An illustration of self assembly of preparation of nanofibers	69
3.6	A schematic of electrospinning setup for the preparation of nanofibers	71
3.7	An illustration of X - Ray Diffraction	75
3.8	Schematic representation of X-Ray Diffractometer	76
3.9	An illustration of energy level transitions between electronic states in a molecule	78
3.10	A simple representation of UV-vis spectrophotometer	79
3.11	Schematic diagram of UV-vis spectroscopy	79
3.12	Fourier Transform Infrared Spectrometer unit	81
3.13	An illustration of FTIR Spectrophotometer	82
3.14	Schematic of Michelson Interferometer employed in FTIR Spectroscopy	83
3.15	Raman Spectroscopy unit	85
3.16	Schematic of Raman Spectrometer	86
3.17	Schematic of TG/DTA system	88
3.18	A typical FESEM system	91
3.19	Coating unit used for FESEM sample preparation (left) and gas cylinder (right)	92

3.20	Schematic diagram of FESEM machine	92
3.21	TEM Detector Unit	94
4.2.1	Electrospinning setup for the preparation of nanofibers	109
4.2.2	Peeling of as-prepared electrospun TiO_2-PVP mat on aluminum foil fig.(b) and collected sample	110
4.3.1	X-ray diffraction patterns of electrospun TiO_2-PVP mat and calcined TiO_2 nanofibers	112
4.3.2	Surface morphology of calcined TiO_2 nanofibers at scale 1µm fig.(a), 200 nm fig.(b), 100 nm fig.(c) and EDX spectra fig.(d)	113
4.3.3	TG/DTA graph of electrospun TiO_2-PVP mat	114
4.3.4	SEM images of TiO_2 nanofibers prepared at voltages 8, 9, 10 and 11 kV	116
4.3.5	SEM images of TiO_2 nanofibers prepared at distances 8, 10, 12 and 14 cm	117
4.3.6	SEM images of TiO_2 nanofibers prepared at flow rates 0.6, 0.8, 1.0 and 1.2 ml/hr	118
4.3.7	SEM images of TiO_2 nanofibers prepared at PVP concentrations 0.6, 0.8, 1.0 and 1.2 g	119
4.3.8	Diameter of TiO_2 nanofibers as a function of applied voltage, distance tip-collector, flow rate and the PVP concentration	120
4.3.9	SEM image fig.(a) and EDS spectra fig.(b) of TiO_2 nanofibers prepared at optimized electrospinning process parameters respectively	123
4.3.10	FTIR spectra of electrospun TiO_2 nanofibers prepared fig.(a) and current density-voltage characteristics of DSSC based on TiO_2 nanofibers fig.(b)	124
5.3.1	XRD patterns and 5.3.1(b),(c),FESEM images of TiO_2 (T) ZnO (Z) nanofibers	138
5.3.2	Surface morphology of TZ11 and TZ12 nanofibers	140
5.3.3	Surface morphology of TZ21 and TZ13 nanofibers	141
5.3.4	TEM micrographs of TZ13 nanofibers fig.(a) and fig.(b), HR-TEM fig.(c) and SAED pattern fig.(d)	143
5.3.5	XRD pattern of TZ13 nanofibers fig.(a) and Uv-vis absorbance of variant composite nanofibers fig.(b)	145

LIST OF TABLES

Table No.	Name of the table.	Page No.
3.1	Capabilities of various nanofibers preparation methods	72
3.2	Merits and demerits of various nanofibers preparation methods	73

LIST OF ABBREVIATIONS

DSSC	Dye-sensitized solar cells
TiO_2	Titanium oxide
ZnO	Zinc oxide
PEE	Personal protective devices
BET	Brunauer-Emmett-Teller
1-D	One Dimension
A	Anatase (A)
R	Rutile (R)
DTG	Differential Thermogravimetry
PVP	Poly-vinylpyrrolidone
TTIP	Titanium tetraisopropoxide
FTO	Fluorine doped tin oxide
EBT	Eriochrome black T dye
DTG	Differential Thermogravimetric
EDS	Energy Dispersive X-ray Spectroscopy
CCD	Charge-coupled device
FESEM	Field Emission Scanning Electron Microscope
FT-IR	Fourier Transform Infrared
T_s	Specimen temperature
T_r	Reference temperature
IR	Infrared Radiation
JCPDS	Joint Committee on Powder Diffraction Standards
SEM	Scanning Electron Microscope
TG	Thermogravimetric
TEM	Transmission Electron Microscopy
XRD	X-ray Powder Diffraction
EDAX	Electron Dispersive Analysis of X-rays
SAED	Selected area electron diffraction

Chapter-1

1.0. Introduction

Nanotechnology is an emerging field which deals with the materials made-up of smaller particles or grains size below 100 nanometer (nm). To understand what nanometer scale is we can consider a human hair which is approximately 1000 times bigger in its diameter. Dealing with the nanomaterials is closer to the atomic scale. For example, 1 nanometer is equivalent to 10 angstroms (A°). Presently, nanomaterials are of great interest by the research community for their potential applications. The demands of nanomaterials are due to their unique optical, electrical, mechanical, chemical etc. characteristics which are due to the high surface-area to volume-ratio and quantum confinement.

Nanomaterials are found to be in the form of nanoparticles, nanowire, nanorods, nanofibers etc. Out of these, one-dimensional nanomaterials are recognized to be promising for supporting the unidirectional transport mechanism to the electrons in electronics and photonics in optoelectronics devices. In this view, one-dimensional structure of nanofibers are being demanded for their typical applications in dye-sensitized solar cells (DSSCs), batteries, super capacitors, sensors (electrical, chemical and biosensors) and so on [1-10]. One of the popular methods for fabricating the nanofibers is an electrospinning method. For the fabrication of nanofibers, the electrospinning approach is an inexpensive one with its easy setup which consists of mainly three parts such as dc power source to induce the electrostatic field, syringe pump to inject the solution, and collector to collect the fiber mat. The properties of prepared nanofibers can be altered by varying the process parameters during the preparation of gel and electrospinning fabrication.

1.1 Background of Electrospinning

The electrospinning process came in existence after patenting the electrospraying in 1902 by C.F.Cooley and W. J. Morton[11,12]. Later, in 1929 A. Farmhals subsequently claimed two patents of electrospinning process[13-14].Later in 1060, the pioneer work reported by G.Taylor brought the attention of scientific community for its importance of electrospinning process [15]. He explored the formation of various shapes of polymer droplet with the applied dc voltage between the metal tip and the collector. Due to his contribution, the formation of cone-shape of polymer was named as Taylor cone. Further, Vonnegut et al. demonstrated the formation of mono dispersed liquid particles by electrical atomization in 1952 [16]. Similarly, Simons patented the simple setup of electrospinning for the preparation of thinner fiber in 1966 and later in 1971 Baumgarten also patented the formation of acrylicfibers using developed electrospinning setup [17,18]. These inventions did not recognize by the scientific community and for the industrial application due to lack of testing equipment to analyze the nanofibers. The attention towards the electrospinning was regained in 90s after the pioneer works published by Reneker group Xia et al. and Wendorff et al. [19-23]. Presently, due to growing interest in nanotechnology, electrospun fibers have been well recognized in the category of one-dimensional nanostructures having a large area-to-volumeratio which is beneficial for the several promising applications. Further, by varying the process parameters, one can alter the properties of nanofibers to a requirement of specific application.

1.2 Electrospinning Process

The principle of electrospinning process lies upon the stretching of the gel or viscous solution under high voltage condition. Due to the strong electrical force, the surface tension of the viscous solution gets controlled and ejects the liquid jet. The typical set-up of electrospinning system is depicted in **figure 1.1**, which mainly consists of high-voltage source, syringe pump and the collector. During the process, the viscous solution is pumped by syringe pump by a regulated flow rate in the presence of applied potential between the collector drum/plate and the metal jet [24].

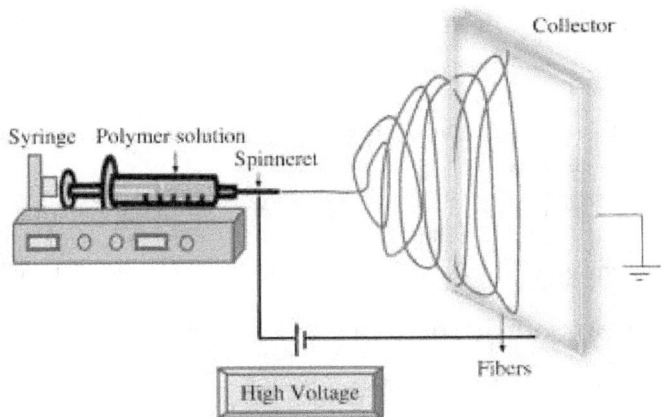

Figure 1.1.The schematic of a typical electrospinning setup.

For the preparation of nanofibers, an optimum applied voltage is needed in order to decrease the surface tension of the polymer based fluid which induced the mutual repulsive forces of the charges. Once this condition is stratified, the solution jet is expelled in the shape of a Taylor sphere. Further, this jet is accelerated to the grounded metal collector placed at distance between 10-30 cm.

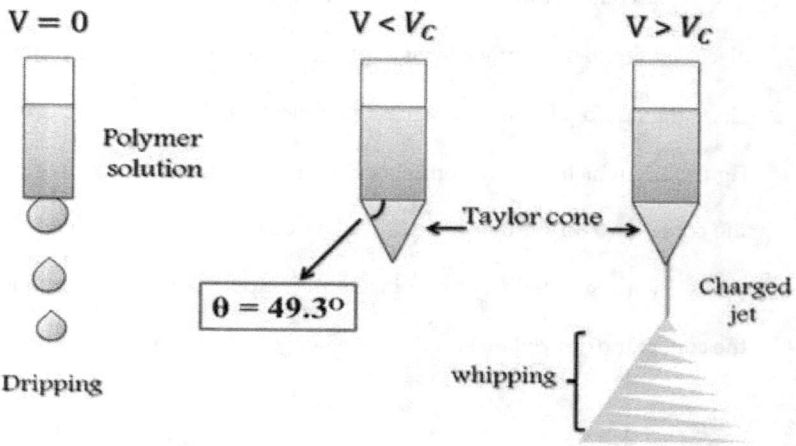

Formation of Taylor cone

Figure 1.2. Formation of Taylor cone by varying the applied the dc potential[25].

During this process, one can observe the instabilities in terms of the bending and axisymmetric flow of the jet. The cause of these is the related to the charges which further, leads the stretching of fibers and reduces diameter when it reaches to the collector. In this process, the solvent gets evaporates and solid fibers are collected on the collector. The formation of Taylor cone with the applied dc potential is illustrated in **figure 1.2**.

For the electrospinning process, the surface tension of the solution is of great importance for jet ejection which needed to be overcome upon the application of critical or threshold voltage (Vc). In the first step, the dropping of polymer solution from the metal tip is usual when no applied voltage (V). In the second step, when the voltage V is applied but lesser than the critical voltage Vc, then the formation of polymer cone is observable. Further, if the applied voltage V is raised to critical voltage then a thin polymer jet will set off and attracts towards the grounded electrode i.e. collector. In other words, after applying the voltage, the electrical charges are gathered on the solution's surface. This generates the mutual repulsive force due to reduced surface tension of the polymer solution therefore, induces the shear stresses in the fluid. The increased electric field yields the hemispherical fluid surface and maintains solution flow in a conical shape i.e. Taylor cone [25].

An electrospinning technique is enormously straightforward and economical one but the process of nanofibers is not so easy. Because to get the Taylor cone of polymer solution for the successful fabrication of nanofibers, the polymer solution has to face various kinds of instabilities which includes Rayleigh instability, axisymmetric and whipping, bending instabilities. An illustration of electrospinning jet path after the applied voltage greater the critical value is depicted in **figure 1.3**.

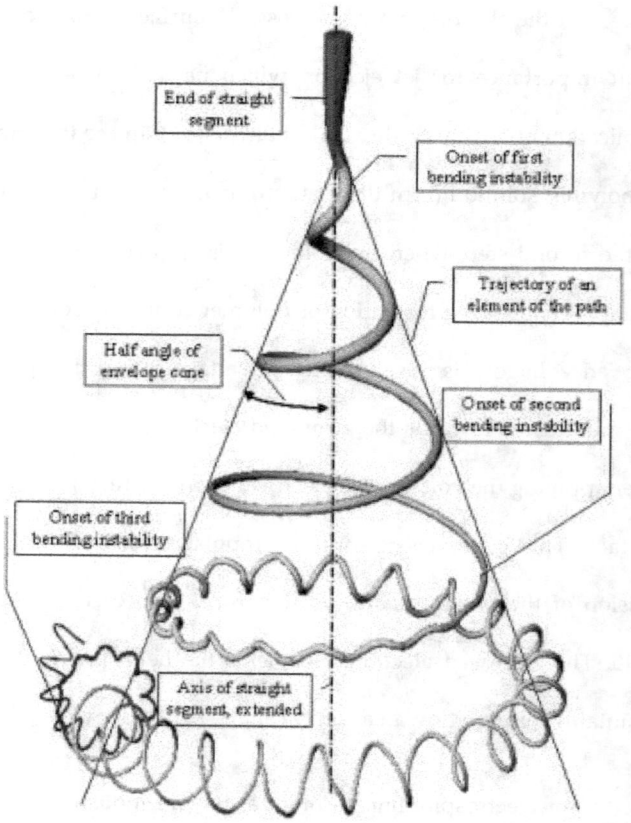

Figure 1.3. Illustration of electrospinning jet path with the applied voltage [25].

Upon the applied voltage, the polymer get is accelerated against the grounded collector while the solvent gets evaporate during the elongation of it due to stretching and whipping. However, before the fibers reach to collector, the polymer solution has to face other steps such as bending, solidification to fabricate the continuous nanofibers. Once the electrospinning process started, we can notice continuous stream of solution following the trajectory of the jet. Further, following the perturbations of the flow we can notice bending of the nanofibers. As this continues the electric charged jet goes under bend and forms coiling-like in three-dimensions. Further, the jet is pulled down and the diameter gets increased due to both bending and elongation mechanisms. However, the coiling process on the collector site reduces the jet diameter which ultimately, enhances the surface area per unit mass of the solution. A further reduced jet diameter deviates the path length of the jet and brings instability. Among all instabilities, the whipping, bending are the important ones and strongly depended on the induced electrostatic charges which are responsible to control the quality of fibers by the way of stretching and thinning during the electrospinning process. Other than these instabilities, others are undesirable which causes the disturbance of jet radius and droplet of the polymer solution.

1.3 Essential Factors for Electrospinning

The morphology of the nanofibers are strongly depends on mainly three factors such as solution properties, control parameters, and ambient environment. The characteristics of solution is influenced by the dipole moment, dielectric constant, crystallinity, glass-transition temperature, molecular weight of the polymer; the polarity, and surface tension of the solvent, viscosity/concentration, elasticity, electrical conductivity; aging of solution etc. Under Control parameter, the flow rate of solution, applied electric field, distance from tip-collector, collector type (plate/drum) and speed and needle related parameters such as its diameter, length, shape geometry like coaxial needle etc. The properties of prepared nanofibers have great influence of ambient environment which includes humidity and temperature of working area.

Using electrospinning setup, we can fabricate of continues nanofibers based on single material (SiO_2, TiO_2, SnO_2, VnO_2 etc.), doped nanofibers (Ag-TiO_2, Cr-TiO_2 etc.), composite nanofibers (TiO_2/ZnO, TiO_2/SnO_2, TiO_2/NiO etc.) and variants morphology nanofibers (porous/core-shell etc.). Even using electrospinning of living bacteria can also be done as reported in the literature [26]. To have the required structural, optical and other characteristics, the solution's properties like viscosity, conductivity and homogeneity are the prominent ones and briefly discussed below.

1.3.1 Viscosity

Viscosity is defined as the flow resistance of the liquid. In general, the intermolecular attraction forces are the key factor for the high or low viscosity of the solution i.e. the stronger the force of attraction in the liquid more the viscosity of the solution or vice-versa. For the electrospinning process without any problem during fiber production, a colloidal, stable and homogeneous solution is highly desirable. According to the reported in literature, the prepared solution should have quite enough viscosity in the range from 0.1-2Pas. This required viscosity of the electrospinning solution may be equal to the viscosity of honey [27-35]. Further, the needed viscosity can be attained by choosing the appropriate polymer of higher molecular weight and optimizing the concentration of polymer during the sol-gel synthesis. In addition, the viscosity is also affected with the choice of solvent which involves its surface tension and the atmospheric temperature during the process.

1.3.2 Conductivity

The conductivity of any substance is defined as the ability to transfer electric current. During electrospinning process, the ionic conductivity of polymer solution is essential for the fabrication of continues electrospun nanofibers. For example, beaded fibers are produced if the ionic conductivity of the polymer solution is very less. This happens as a result of non-homogeneous alignment of charge on the solution surface. To overcome the problem of beaded nanofibers, the dilution of conducting salt in the solution is the choice. However, the use of salt results in contaminated sample which should be eliminated after the electrospinning process is over. The choice of salt is significant for the contaminants free sample. For example, if the solution is polar then ammonium carbonate can be preferred as the conducting salt. The advantage of this salt is that it gets fully decompose intogaseous products (for example, ammonia, water and carbon dioxide) after thermal treatment at moderate temperature. However, calcination of electrospun sample is usually preferred for the removable of inorganic components and also to crystallite it. Though other conducting salts for example, sodium chloride is also the choice but as this is non-volatile compound which cannot be eliminated after heat treatment.

At the same time, the conductivity of the solution should not be too much else this creates short-circuit between the metal tip and the collector electrode. This may yield a disturbance in the sustained steam of the solution which causes the fluctuation in the spinning process and therefore, non-uniform diameter fibers are produced. However, an admissible conductivity of the solution can be tuned with the composition of solution.

1.3.3 Homogeneity

If the elemental composition and the properties are uniform then the solution is said to homogeneous. For the electrospinning process, the required solution may be the mixture of polymer, solvent, precursor, acid, nanoparticles, bacteria etc. Therefore, the homogeneity of the resultant solution must be adequate with the least metastable state. Care should be taken when polar and nonpolar substances are used as these yield precipitation. Therefore, preparation of homogenous solution is a challenging task particular in the electrospinning of inorganic fibers using nanoparticles. The dispersion of inorganic nanoparticles not only depends on the selection of solvent type but also the concentration must be taken into consideration [36-39]. In some cases, a limit of solvent concentration may not be useful for the dilution of a specific polymer. In these circumstances, two considerations must be kept to prepare homogeneous solution for the electrospinning. The first is the use of sufficient quantity of polymer so that a viscous solution can be prepared while second one is the concentration of nanoparticles which should be adequate to get the fiber morphology after the heat treatment.

These conditions are necessary for the preparation of continues nanofibers. To meet this, either, we can choose equal mass concentration of nanoparticles and the polymer concentration or at least 1/3rdconcentration of polymerin order to prohibit the nanofibers morphological disturbance during heat treatment. In this, these factors are important for the tuning of properties of electrospun nanofibers.

1.4 Applications of Electrospun Nanofibers

Electrospun nanofibers are being demanded for their several promising applications. Several industries marketing many products based on electrospun nanofibers due to their large area which felicitates the adequate interaction with the substances. For applications point of view (Figure 1.4), these nanofibers are used for filtration of gas and liquid for the elimination of dust, smoke and particulates turbo machinery, aeroplane cabins, research labs like cleanrooms, water filtration in water filters, oil filtration in engines for efficient performance, dust filtration in gas cartridges, dye-sensitized solar cells, batteries, making of composite membranes, biosensing applications, smart fabrics, and so on [40-53].

1.4.1 Stimuli-Responsive Applications

Nanofibers are useful for the stimuli-responsive application. But the diffusion is a factor which affects the stimuli polymers response [54-56]. A mechanism to get through this issue is the electrospinning of stimuli-responsive polymers. Using this approach, an rapid transport of stimulus was noticed which has been ascribed to the porous morphology of the electrospun nanofibers along with the enhanced surface are to volume ratio which felicitates the smooth transfer mechanism. The parameters which influence the stimulus response are the external field like electrical/optical/thermal/ magnetic one. The thermal stimulus represents the applied temperature which can be studied after preparation of thermo-responsive polymers based fibers. In addition, thermo-responsive nanofibers can be used for shape recovery application with the application of applied temperature [57].

1.4.2 Shape-Memory Applications

These kinds of nanofibers have ability to transform their deformed shape to its primary shape after an external stimulus [58-60]. For such physical transitions, temperature is the stimulus. As the material is reverted to its initial state therefore, we call them shape memory materials. To prepare such shape-memory nanofibers, an electrospinning method is well recognized one. Further, depending upon the choice of polymers the multiple transformation states can be attained by varying the applied temperature. To understand this, one can consider the thermoplastic polyurethane polymer. At room temperature, the shape-memory fiber mat maintains the same shape of the mat and shows transition from original shape to temporary shape and again regains the original shape at room temperature. So what is the matter which brings this kind of thermal transition. The reason of this is related to the polymer chains in the prepared mat which deforms owing to the movement of polymer chains upon the thermal exposure. Further, these polymer chains freeze-off due to storage of entropic energy in the mat when the temperature goes down.

1.4.3 Self-Cleaning Applications

Several materials have been explored for the self-cleaning applications whereas electrospun nonwoven mats are one of promising one for the self-cleaning coatings. Because of their breathable, versatile, and self-supporting ability, the use of nanofibrous mat is beneficial. These are significantly useful to avoid the contamination in medical devices and in manufacturing of the safe medical aprons or cloths [61,62]. By manipulating the nanofibers crosse-section along with chemical formulation, the mat's surface wettability can be controlled to attain the better ability to self-clean. By altering the properties of nanostructures, we can prepare the surface like super hydrophilic and super hydrophobic surface which takes part for the self-cleaning process.

1.4.3.1 Superhydrophilic Surface

If the prepared nanofiber mat is superhydrophilic, water or liquid will scatter to create a thin film on it. In this case, if the mat is inclined, the film will wipe away the dirt. Superhydrophilicity is accomplished mainly by the photocatalytic response of nanofibers. Upon irradiation of ultra-violet light, TiO_2 nanofibers mat showed very small contact angle (<1°). The mat's thickness also influences the water film forming. The flow of water film for an adequately dense mat was a hydrodynamic method with no water beading. The movement of water film, however, required fast balance by surface diffusion for thinner mats. Such nanofiber mats having superhydrophilic surface is often coated on the glass for the self-cleaning process. Wettability is not the only function that is required in this case. Often important is the transparency or low dispersion of the coated glass and this can be influenced largely by the thickness of the mat and its adhesiveness of it. Diethanolamine was used to control the transparency of the nanofiber mat during the electrospinning TiO_2 based fibers [63]. The resulting layer moved from opaque nanofibers to translucent nanoparticles when the volume of diethanolamine was decreased. The use of diethanolamine allowed greater adhesion for the coating over the glass.

1.4.3.2 Superhydrophobic Surface

The nanofiber mat with superhydrophobic nature is actually inspired by the superhydrophobic surfaces existing in nature, like the lotus leaf. With this approach, the surface coating not only has a small surface energy but also shows a graded smoothness[64]. In this way, by rolling the water droplets on the nanofibers-mat, the self-cleaning effect is achieved. There are few specific criteria to be considered in order to achieve a self-cleaning superhydrophobic surface. For example, requirement of large water contact angle, a low water droplet adhesion which felicitates droplets to roll-off below ten degree surface inclination, and a lower dust adhesion than water droplets so that water droplets can deliberately capture the dust particles [65]. In these circumstances, superhydrophobic coating can be made by preparing hydrophobic polymer based fiber using electrospinning process, by alteration rough surface using the low-surface energy content, and roughening of weak-surface energy substance. In general, it is very important for a self-cleaning coating to be durable. Nevertheless, their use is greatly limited by the low mechanical strength of nanofiber-based pads, perhaps that's the main reason that limited goods are being marketed. Manufacturing of self-cleaning nanofibers mats with increased friction and tear resistance is beneficial to maximize their industrial applications viability.

1.4.4 Self-Healing Applications

Engineering materials are often vulnerable to damage when there is crack. The origin of a crack inside a bulk material is often difficult to detect. It's almost impossible to repair the crack. For this reason, it is highly desirable to use self-healing materials that can self-repair damage and so restore their original characteristics. Self-healing, in fact, is an essential characteristic of the biological system that increases living organism's lifespan [66]. Often the nanofibers themselves serve as a curing agent. To realize this, the PCL nanofibers were spread over the epoxy matrix under and a blade cut was marked over that surface. This process was performed at temperature 60 °C however, the molten PCL nanofibers could bind the opposite sides of the crack close and the crack was completely repaired when heated at 80 °C for few minutes [67]. Nevertheless, in this situation, further heat therapy is essential to initiate the process of self-healing. Alternate approaches are established by encapsulating liquid healing agents into distinct morphology based core-shell nanofibers and then integrating nanofibers into a polymer matrix. The healing agents exit the damaged nanofibers and spread to the underlying matrix when a crack migrates via the nanofibers, producing healing materials to restore the crack [68-70]. For this, the monomer and healing agent are the required components for the manufacturing of healing materials but these should not be identified until the composite is destroyed [71].

1.4.5. Living Applications

To order to manufacture "living" nanofibers, even live microorganisms such as bacteria and cells can be introduced during electrospinning process. Such living nanofibers are broadly suitable for the biotechnology based products [72]. A stream containing the bacteria is projected from a suspension developed by dispersing the living microorganisms in a polymer based solution derived by the sol-gel synthesis. Generally speaking, the encapsulated microorganisms are not electrically influenced as the residual charges are dispersed primarily on the jet surface.

Microbes (bacteria/viruses) can be embedded in nanofibers by using the aqueous solution comprising of microbes and a polymer for electrospinning. Typically the bacteria are dissolved in a salt solution or nutrient medium and then blended in the aqueous polymer based solution. According to reported work, the bacteria or viruses incorporated in PVA nanofibers can be preserved even for three months at the maintained temperature -20 and -55°C [73]. The coaxial electrospinning approach can be used to protect the bacteria from the poisonous solvent and expand this process to water-insoluble polymers.

1.4.6 Sensing Applications

Nanofibers are required in the sensors because of their excellent sensitivity and response time due to the porous nature and the enormous surface-to-volume ratio. These are useful with chemical species to identify differences in concentration, such as small molecules of water/glucose, biomolecules, including enzymes or proteins, and gaseous content. A fair response to variations in moisture was demonstrated using 30 wt % $LiCl-TiO_2$ nanofibers[74]. The rapid investigation of physiologically substances has also been explored. Thanks to its function in altering various biological processes, glucose is one of the obvious examples. For rapid identification of glucose, various types of nanofibers were embedded in glucose oxidase or functional nanomaterials, providing an effective mechanism for detection of glucose [75-77].

1.4.7 Energy Applications

It has been shown that the superhydrophobicity nature of PP fibers is useful for the coating on the solar cells to prevent degradation and extend the lifespan of solar cells. The key criterion to attain the hydrophobicity is stated that the size of nanofibers should be maintained below 10 μm. As compared to non-woven mats, nanofibers with reduced diameter showed an enhanced performance [78]. Certain possible energy applications using electrospun nanofibers like electronic circuit architecture, organic, electronics, piezo-electrics, batteries, micro-electro-mechanical systems; and micro/nano/optical/thermal and chemical sensors [79-93].

1.4.8 Filtrationor Separation Applications

Porous nanofiber membranes are becoming more significant for filtering and separating processes in industries. For example, filtering of gas and liquid and separation in electrical devices, batteries etc. [80, 83, 88, 94-105]. Such porous membranes are highly demand for reliable and long-term filtering and separation process [106,107]. Ultrafine fibers offer many advantages compared to traditionally fibers due to high surface energy, porous morphology, and high-strength. In this view, this is promising for the consumption of limited energy for the filtration process. Based on the configuration of the membrane, different fundamental processes can be attained. For instance, the isolation of particles of different sizes, immobilization to enhance precise separation or filtration, use of contact membranes as mediators to facilitate interaction among the two reactive elements[107]. Ultrafine nanofibers can be used for the pretreatment of wastewater before reverse osmosis. This way a lower power consumption and low maintenance can be attained [108].

1.4.9 Textile Applications

Melt electrospun fibers have been recognized for their textile applications such non-wetting surfaces; engineered leather, wiping cloths, military cloths, building composite material reinforcement, chemical protective cloths, agriculture pesticide clothing [84, 93, 95, 97, 103, 105, 109]. The surface chemistry is important for the electrospun webs as an efficient resistant to liquid infiltration into personal protective devices (PPE), as the wetting mechanism relies on the material's surface energy. A weak surface energy plays a major role in repelling the liquid from the cloths. Nevertheless, protective clothing consisting of impermeable fabrics can be dangerous and can cause hyperthermia in high temperature with small evaporation levels. In conjunction with high-barrier efficiency, the breathability of a material in dry and humid conditions for wearer comfort should therefore be considered [110]. PP is the most frequently demanded nonwovens for PPE, since its surface energy is fairly small and chemically inert, light-weight, and expensive. A breathable textile based on PP melted electrospun fiber was demonstrated which enabled water vapor to flow out the surface of the fabric whereas the hydrophobic surface offered the water repellent ability [111].

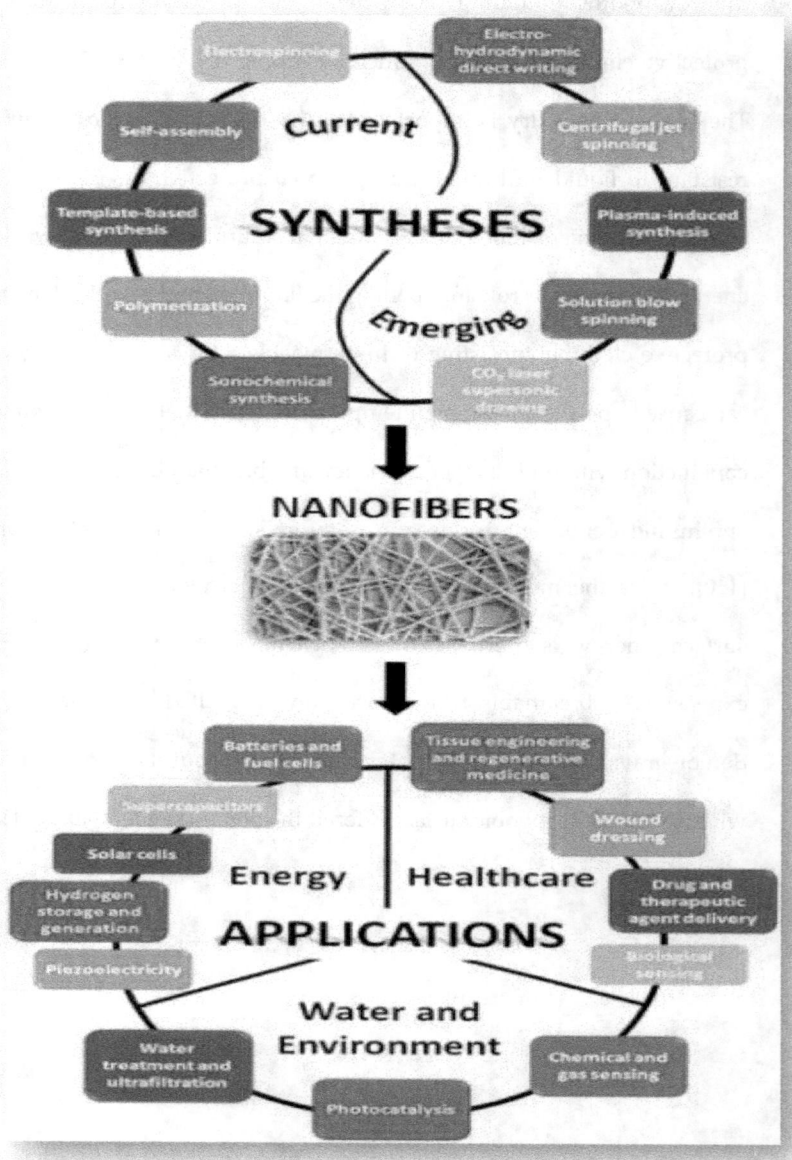

Figure 1.4. Overview of synthesis and applications of nanofibers

References

1. S.M.S. Shahriar, J. Mondal, M.N. Hasan, V. Revuri, D. Y. Lee and Y. K. Lee, Electrospinning Nanofibers for Therapeutics Delivery, Nanomaterials (Basel), Vol. 9(4), 532-32, 2019.

2. Electrospinning and Electrospun Nanofibers: Methods, Materials, and Applications, Jiajia Xue, Tong Wu, Yunqian Dai and Younan Xia, Chem. Rev., Vol. 119(8), 5298-5415, 2019.

3. Koenig, K., Beukenberg, K., Langensiepen, A new prototype melt-electrospinning device for the production of biobased thermoplastic sub-microfibers and nanofibers, Biomater Res, Vol. 23(10), 1-12, 2019.

4. Md Shariful Islam, Bee Chin Ang, Andri Andriyana and Amalina Muhammad Afifi, A review on fabrication of nanofibers via electrospinning and their applications, SN Applied Sciences, Vol.1, 1248-1263, 2019.

5. Dalapathi Gugulothu, Ahmed Barhoum, Raghunandan Nerella, Ramkishan Ajmer and Mikhael Bechelany, Fabrication of Nanofibers: Electrospinning and Non-electrospinning Techniques, Springer, Cham, ISBN 978-3-319-53654-5, 2019.

6. Sezin Sayin, Ali Tufani, Melis Emanet, Giada Graziana Genchi, Zepideh Shemshad, Ece Ozdemir, Gianni Ciofani, Ozlem Sen and Gozde Ozaydin Ince, Electrospun Nanofibers With pH-Responsive Coatings for Control of Release Kinetics, Frontiers in Bioengineering and Biotechnology, Vol.7, 309-12, 2019.

7. Brian S. Chapman, Sumeet R. Mishra and Joseph B. Tracy, Direct electrospinning of titania nanofibers with ethanol, Dalton Trans., Vol. 48, 12822-12827, 2019.

8. Zhuojun Duan, Yingzhou Huang, Dingke Zhang and Shijian Chen, electrospinning Fabricating Au/TiO2 Network-like Nanofbers as Visible Light Activated photocatalyst, Vol. 9, 8008-1-18, 2019.

9. Electrospinning Polymer Nanofibers With Controlled Diameters, IEEE Transactions on Industry Applications, Vol. 55(5), 5239 – 5243, 2019.

10. Mayakrishnan Gopiraman and Ick Soo Kim, Preparation, Characterization, and Applications of Electrospun Carbon Nanofibers and Its Composites, DOI: 10.5772/intechopen.88317, 2019

11. J.F. Cooley, Apparatus for electrically dispersing fluids, US Patent Specification 692631, 1902.

12. W.J. Morton, Method of dispersing fluids, US Patent Specification 705691, 1901.

13. A. Formhals, Verfahren Und Vorrichtung Zur Herstellung von Kuenstlichen Fasern, German Patent 584801, 1929.

14. A. Formhals, Process and Apparatus for Preparing Artificial Threads, U.S. Patent 1975504, 1930.

15. G.Taylor, Electrically Driven Jets. Proc R Soc Lond, A Math Phys Sci, Vol. 313(1515), 453-475, 1969.

16. B. Vonnegut, R.L. Newbauer, Production of monodisperse liquid particles by electrical atomization, J Colloid Sci, Vol.7, 616-622, 1952.

17. H. L. Simons, Process and Apparatus for Producing Patterned Nonwoven Fabrics, US patent 3, 280, 229, 1966.

18. P. K. Baumgarten, Electrostatic spinning of acrylic microfibers, J Colloid Interface Sci, Vol. 36, 71-79, 2014.

19. H. Fong and I. Chun, D, Reneker Beaded nanofibers formed during electrospinning. Polymer (Guildf.), Vol. 40, 4585-4592, 1999.

20. D. H. Reneker and A. L. Yarin, Polymer (Guildf), Electrospinning jets and polymer nanofibers. Polymer (Guildf.), Vol. 49, 2387-2425, 2008.

21. A. L. Yarin, W. Kataphinan and D. H. Reneker, Branching in electrospinning of nanofibers, J. Appl. Phys., Vol. 98(6), 064501-13, 2005.

22. D. Li and Y. Xia, Electrospinning of Nanofibers: Reinventing the Wheel, Adv. Mater., Vol. 16, 1151-1170, 2004.

23. A. Greiner and J. H. Wendorff, Angew. Chemie Int. Ed., Electrospinning: a fascinating method for the preparation of ultrathin fibers, Vol. 46(30), 5670-5703, 2007.

24. Sreeram Ramakrishna, Kazutoshi Fujihara, Wee-Eong Teo, Teik-Cheng Lim and Zuwei Ma, An Introduction to Electrospinning and Nanofibers, World Scientific, ISBN 978-981-256-415-3, 2005.

25. Nabeel Zabar Abed Al-Hazeem, Nanofibers and Electrospinning Method, Novel Nanomaterials-Synthesis and Applications, IntechOpen, DOI: 10.5772/intechopen.72060, 2018.

26. Panikkanvalappil R. Sajanlal, Theruvakkattil S. Sreeprasad, Akshaya K. Samal, and Thalappil Pradeep, Anisotropic nanomaterials: structure, growth, assembly, and functions, Nano reviews, Vol. 2, 5883_1-, 5883_62, 2011.

27. H. Fong, I. Chun and D.Reneker, Beaded nanofibers formed during electrospinning, Polymer, Vol. 40, 4585-4592, 1999.

28. D. Li and Y. Xia, Electrospinning of Nanofibers: Reinventing the Wheel, Adv. Mater., Vol. 16, 1151-1170, 2004.

29. A. Greiner and J. H. Wendorff, Electrospinning: a fascinating method for the preparation of ultrathin fibers, Angew. Chemie Int. Ed., Vol. 46, 5670-5703, 2007.

30. R. Ramaseshan, S. Sundarrajan, R. Jose and S. Ramakrishna, Nanostructured ceramics by electrospinning, J. Appl. Phys., Vol. 102, 111101, 2007.

31. S. Ramakrishna, K. Fujihara, W.-E. Teo, T.-C. Lim, Z. Ma, An Introduction to Electrospinning and Nanofibers, World Scientific Publishing Co. Pte. Ltd., Singapore, 2005.

32. D. Li, J. T. McCann, Y. Xia, M. Marquez, Electrospinning: a simple and versatile technique for producing ceramic nanofibers and nanotubes, J. Am. Ceram. Soc., Vol. 89, 1861–1869, 2006.

33. W.-E. Teo, R. Inai, S. Ramakrishna, Technological advances in electrospinning of nanofibers, Sci. Technol. Adv. Mater., Vol. 12, 013002, 2011.

34. W. Salalha, J. Kuhn, Y. Dror and E. Zussman, Encapsulation of bacteria and viruses in electrospun nanofibers, Nanotechnology, Vol. 17(18), 4675–4681, 2006.

35. Z.-M. Huang, Y.-Z. Zhang, M. Kotaki, S. Ramakrishna, A review on polymer nanofibers by electrospinning and their applications in nanocomposites Compos. Sci. Technol., Vol. 63, 2223–2253, 2003.

36. M. Niederberger, G. Garnweitner, Organic Reaction Pathways in the Nonaqueous Synthesis of Metal Oxide Nanoparticles, Chem. A Eur. J., Vol. 12, 7282–7302, 2006.

37. C. Grote, T. A. Cheema, G. Garnweitner, Comparative Study of Ligand Binding during the Postsynthetic Stabilization of Metal Oxide Nanoparticles, Langmuir, Vol. 28, 14395–14404, 2012.

38. H. Kamiya, M. Iijima, Surface modification and characterization for dispersion stability of inorganic nanometer-scaled particles in liquid media, Sci. Technol. Adv. Mater., 11, 044304, 2010.

39. G. Garnweitner, Polymer Aging, Stabilizers and Amphiphilic Block Copolymers, Nova Science Pub. Inc., New York, ISBN: 978-1-60692-928-5, 2010.

40. Frank Ko, Afaf. El-Aufy, Hoa Lam and Alan G. Macdiarmid, Electrostatically generated nanofibres for wearable electronics, Wearable Electronics and Photonics, Woodhead Publishing in Textiles, 1340, 2005.

41. Z.-M. Huang, Y.-Z. Zhang, M. Kotaki and S. Ramakrishna, A review on polymer nanofibers by electrospinning and their applications in

nanocomposites, Composites Science and Technology, Vol. 63 (15), 2223-2253, 2003.

42. S. Kidoaki, I.K. Kwon and T. Matsuda, Mesoscopic spatial designs of nano-and microfiber meshes for tissue-engineering matrix and scaffold based on newly devised multilayering and mixing electrospinning techniques, Biomaterials, Vol. 26(1):37-46, 2005.

43. John J. Stankus, Jianjun Guan, Kazuro Fujimoto and William R. Wagner, Microintegrating smooth muscle cells into a biodegradable, elastomeric fiber matrix. Biomaterials, Vol. 27(5), 735-744, 2006.

44. Peng Ke, Xiao-Ning Jiao, Xiao-Hui Ge, Wei-Min Xiao and Bin Yu, From macro to micro: structural biomimetic materials by electrospinning, RSC Advances, Vol. 4(75), 39704-39724, 2014.

45. Christopher J. Buchko, Loui C. Chen, Yu Shen and David C. Martin, Processing and microstructural characterization of porous biocompatible protein polymer thin films, Polymer, Vol. 40(26), 7397-7407, 1999.

46. Jayesh Doshi and Darrell H. Reneker, Electrospinning process and applications of electrospun fibers, Journal of Electrostatics, Vol. 35, 151-160, 1995.

47. P. Lu and B. Ding, Applications of electrospun fibers, Recent Patents on Nanotechnology, Vol. 2(3), 169-182, 2008.

48. J. Pelipenko, J. Kristl, B. Jankovic, S. Baumgartner and P. Kocbek, The impact of relative humidity during electrospinning on the morphology and mechanical properties of nanofibers, International Journal of Pharmaceutics, Vol. 456(1), 125-134, 2013.

49. T. J. Sill and H. A. von Recum, Electrospinning: applications in drug delivery and tissue engineering, Biomaterials, Vol. 29(13), 1989-2006, 2008.

50. X.. Zong, K. Kim, D. Fang, S. Ran, B.S. Hsiao and B. Chu, Structure and process relationship of electrospun bioabsorbable nanofiber membranes, Polymer, Vol. 43(16), 4403-4412, 2002.

51. L. Huang, K. Nagapudi, R.P. Apkarian and E. Chaikof, Engineered collagen–PEO nanofibers and fabrics, Journal of Biomaterials Science, Polymer Edition, Vol. 12(9), 979-993, 2001.

52. J. Lannutti, D. Reneker, T. Ma, D. Tomasko and D. Farson, Electrospinning for tissue engineering scaffolds, Materials Science and Engineering C, Vol. 27(3), 504-509, 2007.

53. H. Shao, J. Fang, H. Wang and T. Lin, Effect of electrospinning parameters and polymer concentrations on mechanical-to-electrical energy conversion of randomly-oriented electrospun poly (vinylidene fluoride) nanofiber mats, RSC Advances, Vol. 5(19), 14345-14350, 2015.

54. C. Huang, S. J. Soenen, J. Rejman, B. Lucas, K. Braeckmans, J. Demeester and S. C. De Smedt, Stimuli-Responsive Electrospun Fibers and Their Applications, Chem. Soc. Rev., Vol. 40, 2417−2434, 2011.

55. J. Xie, X. Li and Y. Xia, Putting Electrospun Nanofibers to Work for Biomedical Research, Macromol. Rapid Commun., Vol. 29, 1775–1792, 2008.

56. S. Cao, B. Hu and H. Liu, Synthesis of Ph-Responsive Crosslinked Poly[styrene-co-(maleic sodium anhydride)] and Cellulose Composite Hydrogel Nanofibers by Electrospinning, Polym. Int., Vol. 58, 545–551, 2009.

57. Jiajia Xue, Tong Wu, Yunqian Dai and Younan Xia, Electrospinning and Electrospun Nanofibers: Methods, Materials, and Applications, Chem. Rev., Vol. 119, 5298–5415, 2019.

58. A. Lendlein, Kelch, S. Shape-Memory Polymers. Angew. Chem., Int. Ed., Vol. 41, 2034−2057, 2002.

59. Q. Zhao and Qi, H. J. Xie, T. Recent Progress in Shape Memory Polymer: New Behavior, Enabling Materials, and Mechanistic Understanding, Prog. Polym. Sci., Vol. 49, 79−120, 2015.

60. Y. T. Yao, Xu, Y. C., Wang, B.; Yin, W. L., Lu, H. B. Recent Development in Electrospun Polymer Fiber and Their Composites with Shape Memory Property: A Review, Pigm. Resin Technol., Vol. 47, 47–54, 2018.

61. J. L. Sheng, M. Zhang, Y. Xu, J. Y. Yu, and B. Ding, Tailoring Water-Resistant and Breathable Performance of Polyacrylonitrile Nanofibrous Membranes Modified by Polydimethylsiloxane, ACS Appl. Mater. Interfaces, Vol. 8, 27218–27226, 2016.

62. J. L. Sheng, M. Zhang, W. J. Luo, J. Y. Yu, and B. Ding, Thermally Induced Chemical Cross-Linking Reinforced Fluorinated Polyurethane/ Polyacrylonitrile/Polyvinyl Butyral Nanofibers for Waterproof-Breathable Application, RSC Adv., Vol. 6, 29629–29637, 2016.

63. F. Li, Q. M.Li and H. Kim, Spray Deposition of Electrospun TiO_2 Nanoparticles with Self-Cleaning and Transparent Properties onto Glass. Appl. Surf. Sci., Vol. 276, 390–396, 2013.

64. D. Virovska, D. Paneva, N. Manolova, I. Rashkov and D. Karashanova, Photocatalytic Self-Cleaning Poly(L-lactide) Materials Based on a Hybrid between Nanosized Zinc Oxide and Expanded Graphite or Fullerene, Mater. Sci. Eng., C, Vol. 60, 184–194, 2016.

65. E. Ueda and P. A. Levkin, Emerging Applications of Superhydrophilic-Superhydrophobic Micropatterns, Adv. Mater., Vol. 25, 1234–1247, 2013.

66. K. S. Toohey, N. R. Sottos, Lewis, J. A, Moore and J. S, White, S. R. Self-Healing Materials with Microvascular Networks, Nat. Mater., Vol. 6, 581–585, 2007.

67. X. Luo and P. T. Mather, Shape Memory Assisted Self-Healing Coating, ACS Macro Lett., Vol. 2, 152–156, 2013.

68. C. J. Hansen, W. Wu, K. S. Toohey, N. R. Sottos, S. R. White and J. A. Lewis, Self-Healing Materials with Interpenetrating Microvascular Networks. Adv. Mater. Vol. 21, 4143–4147, 2009.

69. X. F. Wu, Rahman, A. Zhou, Z. Pelot, D. D. Sinha-Ray, S. Chen, B. Payne and A. L Yarin,. Electrospinning Core-Shell Nanofibers for Interfacial Toughening and Self-Healing of Carbon- Fiber/Epoxy Composites, J. Appl. Polym. Sci. Vol. 129, 1383–1393, 2013.

70. M. Lee, W. An, S. Jo, H. S. Yoon and L. Yarin, Self- Healing Nanofiber-Reinforced Polymer Composites: Tensile Testing and Recovery of Mechanical Properties, ACS Appl. Mater. Interfaces, Vol. 7, 19546–19554, 2015.

71. S. L. M. Alexander, L. E. Matolyak and L. T. J. Korley, Intelligent Nanofiber Composites: Dynamic Communication between Materials and Their Environment, Macromol. Mater. Eng. Vol. 302, 1700133, 2017.

72. E. Zussman, Encapsulation of Cells within Electrospun Fibers, Polym. Adv. Technol., Vol. 22, 366–371, 2011.

73. W. Salalha, J. Kuhn, Y. Dror, and E.Zussman, Encapsulation of Bacteria and Viruses in Electrospun Nanofibres, Nanotechnology, Vol. 17, 4675, 2006.

74. Z. Li, H. Zhang, W. Zheng, W. Wang, H. Huang, C. Wang, A. G. MacDiarmid and Y. Wei, Highly Sensitive and Stable Humidity Nanosensors Based on LiCl Doped TiO_2 Electrospun Nanofibers, J. Am. Chem. Soc., Vol. 130, 5036–5037, 2008.

75. X. Ji, Z. Su, P. Wang, G. Ma, and S. Zhang, Hollow Nanofiber Membrane-Based Glucose Testing Strips, Analyst, Vol. 139, 6467–6473, 2014.

76. C. Zhou, L. Xu, J. Song, R. Xing, S. Xu, D. Liu and H. Song, Ultrasensitive Non-enzymatic Glucose Sensor Based on Three- Dimensional Network of ZnO-CuO Hierarchical Nanocomposites by Electrospinning, Sci. Rep., Vol. 4, 7382, 2015.

77. A. Senthamizhan, B. Balusamy, and T. Uyar, Glucose Sensors Based on Electrospun Nanofibers: A Review, Anal. Bioanal. Chem., Vol. 408, 1285–1306, 2016.

78. Daehwan Cho, Huajun Zhoua, Youngjin Cho, Debra Audus and Yong LakJoo, Structural properties and superhydrophobicity of electrospun polypropylene fibers from solution and melt, Polymer, Vol. 51(25), 6005-6012, 2010.

79. Li li Cao, Dun-fan Su and Xiaonong Chen, Fabrication of Multi-Walled Carbon Nanotube/Polypropylene Conductive Fibrous Membranes by Melt Electrospinning, Industrial & Engineering Chemistry Research, Vol. 53, 2308-2317, 2014.

80. Rajkishore Nayak, Rajiv Padhye, Ilias Louis Kyratzis, Yen Bach Truong and Lyndon Arnold, Effect of viscosity and electrical conductivity on the morphology and fiber diameter in melt electrospinning of polypropylene. Textile Research Journal, Vol. 83(6), 606-617, 2013.

81. Rajkishore Nayak, Ilias Louis Kyratzis, Yen Bach Truong, Rajiv Padhye and Lyndon Arnold Nayak, Melt-electrospinning of polypropylene with conductive additives, Journal of Materials Science, Vol. 47(17), 6387-6396, 2012.

82. He Guangqi, Zheng Gaofeng, Zheng Jianyi, Lin Yihong, Wei Jin, Liu Haiyan, Wang Bin and Sun Daoheng, Micro/nano structure written via sheath gas assisted EHD jet., Nano/Micro Engineered and Molecular Systems (NEMS), 8th IEEE International Conference, DOI: 10.1109/NEMS.2013.6559807, 2013.

83. J. C. Singer, R. Giesa and H.-W. Schmidt, Shaping self-assembling small molecules into fibres by melt electrospinning, Soft Matter, Vol. 8(39), 9972-9976, 2012.

84. X. Li, H. Liu, J. Wang and C. Li, Preparation and characterization of poly (ε-caprolactone) nonwoven mats via melt electrospinning, Polymer, Vol. 53(1), 248-253, 2012.

85. Ogata, N., et al., Melt-electrospinning of poly (ethylene terephthalate) and polyalirate, Journal of Applied Polymer Science, Vol. 105(3), 1127-1132, 2007.

86. Yosuke Kadomae, Yasuhide Maruyama, Masataka Sugimoto, Takashi Taniguchi and Kiyohito Koyama, Kadomae, Relation between tacticity and fiber diameter in melt-electrospinning of polypropylene, Fibers and Polymers, Vol. 10(3), 275-279, 2009.

87. Yong Liu, Weimin Yang, Zhiqiang Su and Xiaonong Chen, Morphologies and crystal structures of styrene-acrylonitrile/isotactic polypropylene ultrafine fibers fabricated by melt electrospinning, Polymer Engineering & Science, Vol. 53(12), 2674-2682, 2013.

88. Rangkupan, R. and D.H. Reneker, Electrospinning Process of Molten Polypropylene in Vacuum. Journal of Metals, Materials and Minerals, Vol. 12(2), 81-87, 2003.

89. H. Rajabinejad, R. Khajavi, A. Rashidi, N. Mansouri, and M. E. Yazdanshenas, Recycling of Used Bottle Grade Poly Ethyleneterephthalate to Nanofibers by Melt-electrospinning Method, International Journal of Environmental Research, Vol. 3(4), 663-670, 2009.

90. H. Zhou, T.B. Green, and Y.L. Joo, The thermal effects on electrospinning of polylactic acid melts, Polymer, Vol. 47(21), 7497-7505, 2010.

91. E. Zhmayev, D. Cho, and Y.L. Joo, Nanofibers from gas-assisted polymer melt electrospinning, Polymer, Vol. 51(18), 4140-4144, 2010.

92. E. Zhmayev, H. Zhou, and Y.L. Joo, Modeling of non-isothermal polymer jets in melt electrospinning, Journal of Non-Newtonian Fluid Mechanics, Vol. 153(2), 95-108, 2008.

93. E. Zhmayev, D. Cho, and Y.L. Joo, Electrohydro dynamic quenching in polymer melt electrospinning. Physics of Fluids, Vol. 23,073102, 2011.

94. P.D. Dalton, D. Klee, and M. Moller, Electrospinning with dual collection rings, Polymer, Vol. 46(3), 611-614, 2005.

95. S. Malakhov, Method of manufacturing nonwovens by electrospinning from polymer melts, Fibre Chemistry, Vol. 41(6), 355-359, 2009.
96. Zhou, H., T.B. Green, and Y.L. Joo, The thermal effects on electrospinning of polylactic acid melts, Polymer, Vol. 47(21), 7497-7505, 2006.
97. Paul D.Dalton, Dirk Grafahrend, Kristina Klinkhammer, Doris Klee, and Martin Moller, Electrospinning of polymer melts: Phenomenological observations, Polymer, Vol. 48(23), 6823-6833, 2007.
98. J. Lyons, C. Li, and F. Ko, Melt-electrospinning part I: processing parameters and geometric properties, Polymer, Vol. 45(22), 7597-7603, 2004.
99. E. Zhmayev, D. Cho, and Y.L. Joo, Modeling of melt electrospinning for semi-crystalline polymers, Polymer, 51(1), 274-290, 2010.
100. E. Zhmayev, H. Zhou, and Y.L. Joo, Modeling of non-isothermal polymer jets in melt electrospinning, Journal of Non-Newtonian Fluid Mechanics, Vol. 153(2), 95-108, 2008.
101. J. S. Kim, and D. S. Lee, Thermal properties of electrospun polyesters, Polymer Journal, Vol. 32(7), 616-618, 2000.
102. Rajkishore Nayak, Rajiv Padhye, Ilias Louis Kyratzis, Yen Bach Truong and Lyndon Arnold, Effect of viscosity and electrical conductivity on the morphology and fiber diameter in melt electrospinning of polypropylene. Textile Research Journal, Vol. 83(6), 606-617, 2013.
103. Fengwen Zhao, Yong Liu, Huilin Yuan and Weimin Yang, Orthogonal design study on factors affecting the degradation of polylactic acid fibers of melt electrospinning. Journal of Applied Polymer Science, Vol. 125(4), 2652-2658, 2012.
104. Yosuke Kadomae, Yasuhide Maruyama, Masataka Sugimoto, Takashi Taniguchi and Kiyohito Koyama, Kadomae, Relation between tacticity and fiber diameter in melt-electrospinning of polypropylene, Fibers and Polymers, Vol. 10(3), 275-279, 2009.

105. Dasdemir, M., M. Topalbekiroglu, and A. Demir, Electrospinning of thermoplastic polyurethane microfibers and nanofibers from polymer solution and melt, Journal of Applied Polymer Science, Vol. 127(3), 1901-1908, 2013.

106. Panikkanvalappil R. Sajanlal, Theruvakkattil S. Sreeprasad, Akshaya K. Samal, and Thalappil Pradeep, Anisotropic nanomaterials: structure, growth, assembly, and functions, Nano reviews, Vol. 2, 5883_1-, 5883_62, 2011.

107. Aleksander Gora, Rahul Sahay, Velmurugan Thavasi and Seeram Ramakrishna less, Melt-Electrospun Fibers for Advances in Biomedical Engineering, Clean Energy, Filtration, and Separation, Polymer Reviews, Vol. 51(3), 265-287, 2011.

108. Van der Bruggen, B., M. Mänttäri, and M. Nyström, Drawbacks of applying nanofiltration and how to avoid them: a review, Separation and Purification Technology, Vol. 63(2), 251-263, 2008.

109. Mitchell, S. and J. Sanders, A unique device for controlled electrospinning, Journal of Biomedical Materials Research Part A, Vol. 78(1), 110-120, 2006. 195-196 22, 1

110. Lee, S. and S. Kay Obendorf, Developing protective textile materials as barriers to liquid penetration using melt-electrospinning, Journal of Applied Polymer Science, Vol. 102(4), 3430-3437, 2006.

111. Q. Sun, G.M. Rizvi, C.T. Bellehumeur and P. Gu, Effect of processing conditions on the bonding quality of FDM polymer filaments, Rapid Prototyping Journal, Vol.14(2), 72-80, 2008.

Chapter-2

Literature Review and formulation of the thesis

2.1 Literature Review on Electrospinning Nanofibers

Nanofibers are one of the forms of one-dimensional (1D) nanostructures which are promising for the rapid transport and charge-collection efficiency. Doh et al. presented the optimization of applied voltage and flow rate for the electrospinning fabrication of TiO_2 nanofibers. Furthermore, the photocatalytic activity of TiO_2 nanofibers was tested for the treatment of organic pollutants. The optimized applied voltage and solution flow rate were found to be 0.9 kV/cm and 50 l/min. Furthermore, the TiO_2 nanofibers were coated with the TiO_2 particles to boost the photocatalytic activity. As compared to uncoated TiO_2 nanofibers, the coated nanofibers endorsed the remarkable enhancement in photodegradation [1]. Li et al. reported the photodegradation study using catalysts based on TiO_2 nanofibers by varying their diameter.

Later, optimization of nanofibers diameter was performed by varying the concentration of precursor of titanium. A significant influence of nanofibers diameter was noticed for their photocatalytic activity in presence of rhodamine B dye. An increased photocatalyst activity of TiO_2 nanofibers having their diameter about 200 nm studied in presence of rhodamine B dye. Moreover, the photodecomposition was noticed to be lowered with the decreased diameter to 92 nm. The decreased photocatalytic activity was regarded with the trapping of charges which further resulted in the increased charge recombination and reduced the photodecomposition process [2]. Sadeghi et al. presented the morphological investigation of electrospun TiO_2 nanofibers depending upon the solution's viscosity and electrospinning process parameters. The formation of beaded morphology was examined by using the low viscous solution and further, optimization could yield the preparation of smooth nanofibers of diameter 148 nm [3].

Fu et al. reported the electrospinning fabrication of mesoporous TiO_2 nanofibers with further processed by solvothermal method. The added solvothermal treatment was fruitful for the close packing of TiO_2 grains which felicitated the enhancement in light absorption. Further, photocatalytic activity was improved which was ascribed to the increased light adsorption and charge separation capability as a result of second step solvothermal route [4].

Aligned nanospun fibers have the potential to attain the better optical, chemical, and physical properties. In this contest, Kim et al. explored the electrospinning fabrication of aligned TiO_2 nanofibers by integrating two parallel pieces of aluminum as the collector. They investigated the structural and morphological TiO_2 nanofibers by varying the calcinations temperature. The calcination at temperature 500 ºC produced the single phase anatase-TiO_2 while anatase-TiO_2 was yielded at raised temperature 700 ºC. A usual decrease of fibers diameter was noticed in accordance with the higher thermal treatment. The morphology of the electrospun nanofibers was affected by the Applied voltage i.e. a high voltage process disturbed the alignment of nanofibers. Ultimately, with the proposed method the fabricated nanofibers were reported to have improved charge-transport and so crucial for the fast performing devices [5].

Choi et al. reported the fabrication of well-oriented mesoporous titanium oxide nanofibers. For the preparation of TiO_2 solution, TiO_2 particles were directly dissolved in the viscous polymer based-solution and then electrospinning process was performed. For the comparative study, the remaining solution was directly calcined to prepare the TiO_2 nanoparticles. The prepared TiO_2 nanofibers were about 500 nm in diameter with 20 nm size of nanoparticles having mesopores upto 4 nm. Comparatively, nanofibers endorsed the improved photocatalytic properties as compared to nanoparticles sample. In addition, the production of hydrogen was enhanced with the factor 7 using the electrospun nanofibers as compared the nanoparticles.

The improved photodegradation was assigned to the mesoporous nanofibers with the aligned TiO_2 nanoparticles [6]. Presently, TiO_2 nanofibers are employed to prepare the photoanodes of DSSCs and endorsed the improvement in photovoltaic performance. Lee at al. demonstrated the different approach to improve the performance of DSSCs. Electrospun TiO_2 nanofibers were transferred into TiO_2 nanorods by simple calcination at temperature 400 ºC. DSSC based on nanorods exhibited the enhanced cell efficiency upto 11.09 % as compared to 6.11 % obtained by DSSC based on TiO_2 nanoparticles. The enhanced solar cell was attributed to the nanorods morphology which facilitated the charge transport mechanism. As compared to TiO_2 nanoparticles, the prepared nanorods improved the dye-loading capability and the longer electron lifetime which supported the electron collection as examined by the charge-transport characteristics [7].

A distinct morphology based TiO_2 nanofibers employed for the dye-sensitized solar cells and photocatalytic applications. He et al. reported the co-axial electrospinning preparation of hollow/tubular TiO_2 nanofibers followed by etching in sodium hydroxide aqueous solution to get the porous morphology. The hollow/tubular nanofibers diameter was from 300 to 500 nm and ribbon-like shape morphology their width 200 nm was observed of the porous nanofibers. They observed the high surface area (106.5 m^2/g)of the hollow/tubular nanofibers against the usual electrospun TiO_2 nanofibers having surface area 27.3 m^2/g [8].

Aghasiloo et al. prepared and studied the three variants of nanofibers i.e. porous TiO_2 nanofibers, nanofibers by eliminating glycerin and nanofibers by etching TiO_2/ZnO nanofibers. The BET estimated surface area of porous TiO_2 nanofibers was found to be 128 m^2/g as compared to surface area 60-63 m^2/g of later two cases. Accordingly, the photodegradation of methylene blue dye was found to be enhanced as compared to others [9]. Lou et al. presented the study of electrospun PVDF/titanium oxide nanofibers for the photodegradation of RhB dye. By varying the TiO_2 nanofibers concentration, they have concluded that the 20 % concentration of TiO_2 nanofibers were fruitful for the 80 % degradation of RhB within 6 h. They employed the visible light source to study the photocatalyst activity of electrospun nanofibers [10]. Im et al. studied the preparation of TiO_2 nanofibers using PAN and explored the photocatalyst study. The XRD investigation endorsed pure anatase-TiO_2phase and further, the study was explored for their application as the photocatalyst [11].

Chuangchote et al. proposed the easy approach for the fabrication of TiO_2 nanofibers. They used PVP, titanium (IV) butoxide, and acetylacetone materials in the sol-gel process. The diameter of the prepared TiO_2 nanofibers were from 260-355 nm whereas the nanofibrils bundle size was from 20-25 nm. Electrospun nanofibers used as the photocatalysts for hydrogen evolution and demonstrated the enhanced performance as compared to nanoparticles and the samples treated with hydrothermal process.

The improved results are assigned to the improved crystallinity and the increased surface area [12]. Doped TiO_2 nanofibers are also being demanded for the several applications. Tyagi et al. studied the preparation and characterization of tantalum (Ta) doped TiO_2 nanofibers for the supercapacitor application. The 2% Ta doped nanofibers brought the enhancement in charge storage capability which was assigned to the better electrical conductivity which ultimately affected the charge transport mechanism. The supercapacitor prepared using the 2% Ta doped sample showed the specific capacitance 81 F g^{-1} at current density of 0.1 A g^{-1} with its cycling stability of charge-discharge to 5000 [13].

Wang et al. demonstrated the photodegradation study by using gold doped-TiO$_2$ nanofibers fabricated by the simple electrospinning method. The improved photocatalytic activity was assigned to the generated Schottky-barrier at the interface of gold-TiO$_2$ which rendered the carrier recombination and hot-electron induction [14]. Furthermore, Han et al. reported the fabrication of silver doped-TiO$_2$ nanofibers by varying the silver concentration. An increased doping concentration evidenced the corresponding increase in the nanofibers diameter along a decrease in photoluminescence intensity. These fibers were used for the antibacterial study with the pathogenic bacteria and showed the unusual result in comparison to TiO$_2$ nanofibers [15].

Nanomaterials have been demanded for gas/chemical sensing applications. Kim et al. employed the electrospun TiO$_2$ fiber mat on the platinum electrode for gas sensing application. The prepared TiO$_2$ nanostructures was single-crystal in nature of anatase phase The device was tested for the sensing ability of NO$_2$ in atmospheric condition which evidenced the usual sensitivity with its detection limit upto 1 ppb [16]. In another approach, Zhang at al. employed the electrospun TiO$_2$ nanofibers for biological application. Nanofibers based substrate showed the enhanced encapsulation of circulating tumor cells [17].

Deniz et al. presented the two steps fabrication of TiO_2 nanofibers. In first step, they electrospun TiO_2 nanofibers were calcined. Again the calcined nanofibers were dispersed in various polymer solutions and refabricated by electrospinning process. By doing so, the morphology of the finally fabricated nanofibers was found distinct (like fiber-in-fiber) and furthermore, photocatalytic activity was explored [18]. The transformation of insulator TiO_2 nanofibers to conductive one was reported by Wang et al. They have used potassium hydroxide which resulted in the reduced resistivity. With this treatment, TiO_2 nanofibers was found suitable for the preparation of supercapacitor and the capacitance was noticed to be enhance about 1500 times the conventional TiO_2 nanofibers [19].

Guo et al. studied the electrospun ZnO nanofibers by using zinc acetate as the source of zinc and PVP as the polymer. The prepared ZnO nanofibers exhibited the polycrystalline nature with the wurtzite structure. The surface morphology investigation endorsed the preparation of uniform and long nanofibers with their diameter from 90-240 nm further, they performed the electrical property and the result is explored [20].

Saidin et al. studied the morphology of ZnO nanofibers by varying the applied voltage. They noticed the variation of nanofibers diameter in accordance with the applied voltage and finally, continuous growth of ZnO nanofibers was attained after the optimization process [21].

Imran et al. presented the investigation of ZnO nanofibers prepared by electrospinning method. With the decreased concentration of zinc acetate precursor, the beaded morphology of nanofibers was noticed. By thermo-gravimetric investigation, the calcination temperature 480 ºC was examined. FTIR vibration peak observed at wavenumber 472 cm-1 was assigned to ZnO nanofibers while the formation of hexagonal wurtzite structure was evidenced by the XRD analysis. Furthermore, by varying the zinc acetate concentration, various samples were prepared and explored for optoelectronic device applications[22]. Similarly Mauro et al. presented the fabrication of ZnO nanofibers and explored the morphology after calcination at various temperatures from 350-650 °C. Further, they explored the structural, optical, and morphological properties and suggested for the photodegradation studies [23].

Wei et al. demonstrated the electrospun formation of solid and hollow ZnO nanofibers for the gas sensing application. Various measurements were carried out before performing the gas sensing application of ZnO nanofibers and hollow fibers exhibited the superior acetone sensing at 220 ºC with its better stability over the solid nanofibers [24].

Similarly, Huang et al. reported the various properties of zinc oxide and later employed for the gas sensing application. They noticed the unusual sensing ability of sensor for the ethanol with operating temperature 300°C. The obtained response and recovery time for the fabricated sensor were reasonable and found promising for the advance sensors [25].

In another approach, Choi et al. studied the electronic transport and gas sensing characteristics of ZnO nanofibers prepared by two-step method. Initially, the polymer nanofibers were fabricated by electrospinning and later zinc oxide sputtering was done on these nanofibers. By these processes, the prepared hollow ZnO nanofibers were endorsed with the grain size about 23 nm. The electronic property analysis showed the depletion of charge carriers on the surfaces (in and out) of the hollow nanofibers due to the localized electron at O-adions which restricted the current flow via core medium [26].

Park et al. presented the electrical property of electrospun ZnO nanofibers. They noticed the improvement in the crystallinity of nanofibers with respect to the annealing temperature in a controlled manner and the diameters while the diameter was varied from 35-100. The electrical property showed the annealing temperature as the dominant factor to attain the optimal initialization energy to induce the conduction through ZnO nanofibers. [27].

Wu et al. employed the ZnO nanofibers in poly(3-hexylthiophene) based solar device. The efficiency was found influenced with the variation of thickness of fibrous layer. A thicker layer of nanofibers shows the decreased performance due to the exhibited defect state which further decreased the lifetime of the electrons [28]. Kim et al. used the ZnO nanofibers for DSSCs application and reported the distinct morphology of the prepared nanofibers by preferring the hot-press of as-prepared fibers prior to calcination. This resulted in the twisted morphology of the nanofibers and endorsed the enhanced efficiency of DSSC [29].

Li et al. reported the hybrid structured of hydrothermally sensitized ZnO nanorods on electrospun ZnO nanofibers with the controlled growth density of ZnO nanorods. The prepared hybrid nanostructure was encapsulated in-between the two metal/Au while electrodes and the current-time response by switching on and off of UV light. In addition, visible light sensitivity was noticed after sensitizing the hybrid nanostructures in ruthenium dye. Moreover, the sensitivity of UV and visible light was found to be as the function of density of nanorods prepared on the ZnO nanofibers [30].

Zhu et al. presented the investigation of UV sensors based on well-oriented electrospun ZnO nanofibers. The prepared sensor was found to be sensitive towards the UV light exposure due to the gathering of the elections at the grain boundaries. Further, possible improvement of the device performance and future implementations were explored [31]. Wu et al. demonstrated the fabrication of field-effect transistor using ZnO nanofiber prepared by electrospinning process.

To fabricate the device, two parallel silver plates were placed over the doped p-type silicon substrate covered with an oxide layer. For the electrical connections, these collector plates were grounded with a copper wire. After electrospinning process for just 5 seconds, the substrate was calcined at temperature 500 ºC for few hours. Finally, electrical measurement was performed which endorsed the characteristic of n-type semiconductor of ZnO nanofibers. This approach was suggested to be as the inexpensive fabrication approach of electrical and optoelectronic devices based on one-dimensional nanostructures [32].

Sui et al. demonstrated the white light emission from ZnO nanofibers fabricated by the simple electrospinning process. Various investigations were carried out to analyze the optical, structural, and morphological properties of as-spun nanofibers. Photoluminescence investigation exhibited the broad peak in the UV-visible spectrum. They suggested the application of ZnO nanofibers for the white light light-emitting devices [33].

Stafiniak et al. employed the electrospun ZnO nanofibers for the biosensor fabrication. They proposed a technique to fabricate the AlNx film for the further processing on it by adopting the reactive magnetron sputtering method. This idea was useful for the fabrication of ZnO nanofibers sensor with the use of photolithography process. The prepared sample was intensively studied for it various properties like surface morphology and topography. Further, they explored the current-voltage I–V analysis to evaluate the electrical and sensing properties [34].

Both TiO_2 and ZnO are the equally demanded materials in several applications such as batteries, sensors, photocatalytic/water splitting, dye-sensitized solar cells etc. Lee at al. presented the preparation and application of TiO_2/ZnO nanofibers for their photocatalytic activity. They presented the comparative study of photodegradation using TiO_2 and composite TiO_2/ZnO nanofibers as the catalysts. The composite nanofibers showed the enhanced photodegradation of methylene blue dye with 96.4% reduced absorbance [35].

Liu et al. employed the electrospun TiO_2/ZnO nanofibers and studied the photodegradation of RhB and phenol. They observed 100% photodegradation of Rh B and 85% of phenol. The enhanced decomposition of RhB and phenol was attributed to the mixing of ZnO in composite nanofibers. They found 15.76 wt% of ZnO mixed TiO_2/ZnO nanofibers as the optimized wt% to have 100 % degradation of dyes [36].

Hwang et al. studied the dynamic antimicrobial activity of electrospun ZnO/TiO_2 composite nanofibers against gram-negative Escherichia coli and gram-positive Staphylococcus aureus [37]. Wang et al. presented the ammonia gas detection using TiO_2/ZnO nanofibers mat after coating with highly porous polypyrrole. Using the prepared sample, they found the fast sensing response along with excellent sensitivity [38].

Fragala et al. studied the cathodeoluminescence of Zn-doped TiO_2 (core)/ZnO shell nanofibers. After doping with Zn in TiO_2, they found phase transition into rutile from the anatase. Cathodoluminescence study of core-shell nanofibers was performed and noticed the remarkable emission peaks of UV and visible wavelength. This was due to the excellent surface area of the doped-TiO_2 core and ZnO shell nanofibers [39].

Park et al. investigated a novel method to fabricated TiO_2–ZnO core-shell nanofibers. TO do so, they first preferred the electrospinning process to fabricate TiO_2 nanofibers to serve as core and later ZnO deposition to serve as shell was carried by atomic layer deposition technique. Finally, O_2 sensing ability of electrospun TiO_2–ZnO core-shell nanofibers was performed and they found the improved sensitivity and better repeatability of the sensor [40]. In another work, Du et al. employed the TiO_2/ZnO core–sheath nanofibers as the photoanode materials for DSSCs. They noticed DSSC based on core-sheath nanofibers showed the enhanced efficiency against the bare TiO_2 nanofibers based one device. The improved efficiency of the device was assigned to the enhanced light harvesting and charge collection efficiency [41].

Jun et al. demonstrated the humidity sensors using ZnO and TiO_2 nanofibers prepared by electrospinning technique. They performed the sensitivity of humidity sensors by using one layer of single materials and two sequential layers of ZnO & TiO_2 nanofibers and found excellent performance in the later case [42].

Wang et al. presented an approach to prepare porous TiO_2/ZnO composite nanofibers whereas $ZnCl_2$ was used to maintain the porosity. They investigated the structural and morphological properties and further studied for N_2 adsorption/desorption isotherm. By studying the surface photovoltage spectroscopy and photodegradation performance of porous TiO_2/ZnO composite nanofibers they noticed excellent result in regard to pure TiO_2 or ZnO nanofibers [43].

Pei et al. explored the electrospinning preparation of TiO_2/ZnO nanofibers the by varying the calcination temperature in order to change the ratio of anatase and rutile ratio for their impact on the photocatalytic activity. With the optimization process, they could demonstrate the calcination temperature 650 °C and anatase/rutile ratio 48:52 were the optimal and effective for the investigation of photodegradation of RhB [44].

Wang et al. employed the combined techniques of electrospinning and hydrothermal to fabricate the TiO_2/ZnO composite nanostructures. Their investigations revealed the better sensing ability of TiO_2/ZnO composite fiber based sensor with the response of 15.7 to 100 ppm of ethanol at 280 µC against to TiO_2 and ZnO nanofibers. This result was found associated to the heterojunction in-between TiO_2 and ZnO and led the fast movement of charge carriers in an effective way [45].

In a similar way, Kanjwal et al. employed ZnO/TiO_2 nanofibers fabricated by both electrospinning and hydrothermal processes. Among pristine-ZnO and TiO_2 nanofibers, the composite nanofibers were found efficient with their boosted photocatalytic activity. The time duration of methyl red and rhodamine B dyes were 90 and 105 min respectively whereas

3 h photodegradation time was noted for other catalysts [46]. Chun et al. investigated the photodegradation study of RhB dye using TiO_2/ZnO nanofibers as the catalyst. They optimized the proportion ratio of anatase and rutile-TiO_2 by preferring the calcination at various temperatures. Doing so, the photocatalytic activity of nanofibers was found to be improved by degrading the rhodamine B dye with the sample calcined at temperature 650 ºC having ratio 48/52 % of anatase/rutile [47].

Yar et al. presented the photodegradation study using polyacrylonitrile nanofibers decorated with the TiO_2, ZnO and TiO_2/ZnO composite nanoparticles. These nanofibers endorsed the enhanced photocatalytic activity as compared to bare PAN [48]. Similarly, Li et al. investigated the TiO_2/ZnO composite nanofibers and evidenced the improved photocatalytic activity. They also performed the recycling of photodegradation study and found the good the stability of the catalyst [49].

Araujo et al. presented the photodegradation study of rhodamine B dye using TiO_2 nanofibers capped with ZnO nanorods. Morphology investigation revealed the three-dimensional geometry of hexagonal wurtzite ZnO nanorods prepared on TiO_2 nanofibers. The use of these nanostructures as the photocatalyst showed nearly 90 % degradation of rhodamine B dye within 70 [50].

Lotus et al. fabricated the TiO_2/ZnO hybrid nanofibers with their diameter from 50-150 nm using electrospinning process and explored the various investigations. UV-vis absorption result exhibited the two peaks associated with the energy gap values 3.0 and 3.5 eV of TiO_2.

The prepared sample exhibited both anatase and rutile phases of TiO_2 along with wurtzite phase of ZnO. Further, they concluded the electrospinning approach as the better techniques to attain the desired properties of composite TiO_2/ZnO nanofibers [51]. Liu et al. prepared the core/shell (ZnO/TiO_2) nanofibers for the photocatalytic application. XRD study exhibited the anatase and rutile phases of the TiO_2 while hexagonal wurtzite phase was noticed corresponding for the ZnO. Comparatively, the core/shell (ZnO/TiO_2) nanofibers showed the red-shift regarded the less activation energy requirement in contrast to individual ZnO and TiO_2 nanofibers. As a result, the ZnO/TiO_2 core-shell nanofibers evidenced the enhanced photocatalytic activity along with their recyclability [52].

To study the photocatalytic activity, Li et al. presented a distinct experimental process to prepare the heterojunction ZnO/TiO_2 composite fibers. The approach was zinc plating on the electrospun TiO_2 nanofibers and then the heat treatment. As a result of these processes, the photocatalytic activity of ZnO/TiO_2 nanofibers was noticed to be reasonably betteras compared to pure-TiO_2 nanofibers [53]. In another work, Hwang et al. investigated the ZnO/TiO_2 hybrid nanofibers and explored for their antimicrobial activity.

A promising antimicrobial activity was investigated in the presence of gram-negative Escherichia coli and gram-positive Staphylococcus aureus under ultra-violet exposure and in the dark [54]. Liu et al. employed the mixing of ZnO powder with the electrospun TiO_2/ZnO nanofibers calcined at different temperatures. To improve the photocatalytic property, the ZnO was blended in TiO_2/ZnO nanofibers. Further, they observed the degradation of RhB and phenol upto 100 % and 85 % respectively by using ZnO wt % 15.76 [55].

Park et al. studied the various characteristics of ZnO nanofibers prepared by electrospinning process. The crystallinity of the nanofibers was found to be dependent on the calcination temperature and further, their study demonstrated the significant role of calcination temperature in order to attain a good electrical conductivity. Finally, they performed the carbon dioxide gas sensing ability and reported the reliability of the measurement [56].

2.2 Formulation of the Problem and Organization of Thesis

Through a literature review, the electrospinning method is concluded to be an easy and inexpensive method for the preparation of variety of nanofibers for their potential application in DSSCs, batteries, fabrics, photocatalytic activity, water filtration and many more. It is also significant to mention that an electrospinning approach is reliable in terms of the reproducibility of electrospun nanofibers.

Several literatures reported the fabrication of TiO_2, ZnO, SiO_2, VnO_2, SnO_2, $BaTiO_3$, composite TiO_2/ZnO, TiO_2/SnO_2 nanofibers etc. and further, explored for their applications. However, the process parameters play significant role to achieve the desired structural, optical, chemical, and morphological properties of TiO_2, ZnO or composite TiO_2/ZnO nanofibers. Electrospun fiber's morphology is mainly dominated by the process parameters which include dc voltage, the flow rate of solution, jet tip-collector distance, polymer concentration and the atmospheric conditions like temperature or humidity.

The nanofiber's diameter is limited by the change in applied dc voltage and the distance from tip-collector. Nonetheless, the formation of nanofibers is possible if the viscosity of the solution is quite moderate else beads or particles are formed via electros praying rather than by an electrospinning process. Likewise, the decreased flow rate of the solvent and the polymer content creates thinner nanofibers. Moreover, the atmospheric conditions such as humidity and temperature have a vital role for the preparation of continuous and defect-free nanofibers.

By controlling the applied voltage, the distance tip-collector, the flow rate, and the polymer concentration, the diameter of the electrospun nanofibers can be controlled. In this way, an electrospinning method is an easy and cheap method, while the quality of nanofibers can be attained by selecting the appropriate process parameters. Thus by understanding the importance of process parameters, this doctoral study addresses the systematic optimization process of TiO_2 nanofibers by considering the applied dc voltage, solution flow rate, tip-collector distance and polymer concentration.

In our best knowledge, such broad study has not been reported earlier. Throughout the fabrication of TiO_2 nanofibers in terms of above listed parameters, the temperature, and the humidity was maintained to 25 ºC and 35 % respectively. After optimizing these parameters, the prepared TiO_2 nanofibers are studied for their structural, optical, and morphological properties in the view of applications in DSSCs. Further, we have fabricated the composite TiO_2/ZnO nanofibers and explored the various investigations for the application as the photocatalysts.

The present thesis contains six chapters concerned with the fabrication and characterization of electrospun TiO_2, ZnO and composite TiO_2/ZnO nanofibers carried out during the proposed doctoral study. This thesis is organized as follows:

Chapter 1

This chapter gives the overview of electrospinning method which begins with brief history. The following chapter explores the electrospinning approach with the essential parameters of electrospinning process which includes solution's viscosity, conductivity, homogeneity and so on. Further, it explores the understanding of various applications of electrospun nanofibers for the stimuli-responsive, shape-memory, self-cleaning, self-healing, sensing, energy, filtration, textile and living applications.

Chapter 2

This chapter focuses on the literature survey on the fabrication, characterization and applications of TiO_2, ZnO and TiO_2/ZnO composite electrospun nanofibers. Further, this chapter emphasizes the formulation of the problem.

Chapter 3

This chapter explores the various fabrication techniques for the nanofibers which comprises template synthesis, drawing method, phase separation method, self-assembly method, and electrospinning approach. The following chapter presents the various characterization techniques used to investigate the structural, optical, chemical, thermal and morphological properties. It includes X-ray diffraction (XRD), UV-vis Spectroscopy, Fourier-Transform Infrared spectroscopy (FTIR), Raman Spectroscopy (RS), Thermogravimetry/Differential Thermal Analysis (TG/DTA), Field-mission Scanning Electron Microscopy (FESEM), Transmission Electron Microscopy (TEM), and Energy Dispersive Spectroscopy (EDS/EDAX).

Chapter 4

This chapter describes the experimental work carried out on process parameters dependent investigation of electrospun TiO_2 nanofibers using X-ray diffraction (XRD), Fourier-Transform Infrared spectroscopy (FTIR), Thermogravimetry/Differential Thermal Analysis (TG/DTA), Field-mission Scanning Electron Microscopy (FESEM), and Energy Dispersive Spectroscopy (EDS/EDAX). The following chapter covers the influence of applied dc voltage, the flow rate of solution or gel, tip-collector distance and the polymer concentration on the diameter of electrospun TiO_2 nanofibers. Finally, dye-sensitized solar cell was fabricated and studied.

Chapter 5

This chapter focuses the experimental work on ZnO and TiO_2/ZnO composite nanofibers for photocatalytic applications. Initially, this chapter describes the fabrication and investigation of pure TiO_2, ZnO and TiO_2/ZnO composite nanofibers using X-ray diffraction (XRD), UV-vis Spectroscopy, Field-mission Scanning Electron Microscopy (FESEM), and Transmission Electron Microscopy (TEM). Later, photocatalytic activity of the various prepared nanofibers was studied.

Chapter 6

This chapter concludes the doctoral work and discusses the future remarks of the work.

References

1. Seok Joo Doh , Cham Kim, Se Geun Lee, Sung Jun Lee, Hoyoung Kim, Development of photocatalytic TiO_2 nanofibers by electrospinning and its application to degradation of dye pollutants, Journal of Hazardous Materials 154 (2008) 118-127.

2. Heping Li, Wei Zhang, Bin Li, and Wei Pan, Diameter-Dependent Photocatalytic Activity of Electrospun TiO_2 Nanofiber, J. Am. Ceram. Soc., 93 [9] 2503-2506 (2010).

3. Soraya Mirmohammad Sadeghi, Mohammadreza Vaezi, Asghar Kazemzadeh and Roghayeh Jamjah, Morphology enhancement of TiO_2/PVP composite nanofibers based on solution viscosity and processing parameters of electrospinning method, 2018 J. Appl. Polym. Sci., Vol. 135, 46337, 11pg, (2018).

4. Junwei Fu, Shaowen Cao, Jiaguo Yu, Jingxiang Low and Yongpeng Lei, Enhanced photocatalytic CO-reduction activity of electrospun mesoporous TiO_2 nanofibers by solvothermal treatment, Dalton Trans, Vol. 43, 9158-9165 (2014).

5. Jae-Hun Kim, Jae-Hyoung Lee, Jin-Young Kim and Sang Sub Kim, Synthesis of Aligned TiO_2 Nanofibers Using Electrospinning, Appl. Sci. 2018, 8, 309.

6. Kovtyukhova, N.I., Mallouk, T.E., Nanowires as Building Blocks for Self-Assembling Logic and Memory Circuits. Chem. Eur. J. 8, 4354-4363 (2002).

7. Byung Hong Lee,Mi Yeon Song, Sung-Yeon Jang, Seong Mu Jo,† Seong-Yeop Kwak, and Dong Young Kim, Charge Transport Characteristics of High Efficiency Dye-Sensitized Solar Cells Based on Electrospun TiO_2 Nanorod Photoelectrodes, J. Phys. Chem. C 2009, 113, 21453-21457.

8. Guangfei He, Yibing Cai, Yong Zhao, Xiaoxu Wang, Chuilin Lai, Min Xi, Hao Fong and Zhengtao Zhu, Electrospun anatase-phase TiO_2 nanofibers with different morphological structures and specific surface areas, Journal of Colloid and Interface Science, Vol. 398, 103–111 (2013).

9. Peyman Aghasiloo, Maryam Yousefzadeha, Masoud Latifia, Rajan, Highly porous TiO_2 nanofibers by humid-electrospinning with enhanced photocatalytic properties, Journal of Alloys and Compounds, Volume 790, 25 June 2019, Pages 257-265

10. Lihua Lou, Jilong Wang, Yong Joon Lee, and Seshadri S. Ramkumar, Visible Light Photocatalytic Functional TiO_2/PVDF Nanofibers for Dye Pollutant Degradation, Part. Part. Syst. Charact. 2019, 1900091.

11. Qian Tang, Xianfeng Menga, Zhiying Wang, Jianwei Zhou, HuaTang, One-step electrospinning synthesis of TiO_2/g-C3N4nanofibers with enhanced photocatalytic properties, Applied Surface Science, Vol. 430, 253-262 (2018).

12. Jalili, R., Morshed, M., Ravandi, S.A.H. Fundamental Parameters Affecting Electrospinning of PAN Nanofibers as Uniaxially Aligned Fibers. J. Appl. Polym. Sci., 101, 4350–4357 (2006).

13. Ankit Tyagi, Narendra Singh, Yogesh Sharm and Raju Kumar Gupta, Improved supercapacitive performance in electrospun TiO_2 nanofibers through Ta-doping for electrochemical capacitor applications, Volume 325, 15 March 2019, Pages 33-40

14. Tao Wang, Yu Zhang, Yong Wang, Jinxin Wei, Ming Zhou, Zhengmei Zhang and Qi Chen, One-Step Electrospinning Method to Prepare Gold Decorated on TiO_2 Nanofibers with Enhanced Photocatalytic Activity, Journal of Nanoscience and Nanotechnology, 18, 3176–3184 (2018).

15. M. A. Kudhier, R. S. Sabry, Y. K. Al-Haidarie and M. F. AL-Marjani, Significantly enhanced antibacterial activity of Ag-doped TiO_2 nanofibers synthesized by electrospinning, Materials Technology, DOI: 10.1080/ 10667857.2017.13967 (2017).

16. Il-Doo Kim, Avner Rothschild, Byong Hong Lee, Dong Young Kim, Seong Mu Jo, and Harry L. Tuller, Ultrasensitive Chemiresistors Based on Electrospun TiO_2 Nanofibers, Nano Lett., Vol. 6, No. 9, 2006.

17. Nangang Zhang, Yuliang Deng, Qidong Tai, Boran Cheng, Libo Zhao, Qinglin Shen, Rongxiang He, Longye Hong, Wei Liu, Shishang Guo, Kan Liu, Hisan-Rong Tseng, Bin Xiong, and Xing-Zhong Zhao, Electrospun TiO_2 Nanofi ber-Based Cell Capture Assay for Detecting Circulating Tumor Cells from Colorectal and Gastric Cancer Patients, Adv. Mater. 2012, DOI: 10.1002/adma.201200155.

18. Ali E. Deniz, Asli Celebioglu, Fatma Kayaci, Tamer Uyar, Electrospun polymeric nanofibrous composites containing TiO_2 short nanofibers, Materials Chemistry and Physics 129 (2011) 701–704.

19. X. He, C.P.Yang, G.L.Zhang, D.W.Shi, Q.A.Huang, H.B.Xiao, Y.Liu and R.Xiong, Supercapacitor of TiO_2 nanofibers by electrospinning and KOH treatment, Materials and Design, 106, 74–80 (2016).

20. Ji Guo, Yang Song, Dairong Chen, and Xiuling Jiao, Fabrication of ZnO Nanofibers by Electrospinning and Electrical Properties of a Single Nanofiber, Journal of Dispersion Science and Technology, 31:684–689, 2010.

21. Nur Ubaidah SAIDIN1, Thye Foo CHOO, Kuan Ying KOK, Mohd Reusmaazran YUSOF1,d and Inn Khuan NG, Fabrication and Characterization of ZnO Nanofibers by Electrospinning, Materials Science Forum, Vol. 888, pp 309-313, 2016.

22. Muhammad Imran, Sajjad Haider, Kaleem Ahmad, Asif Mahmood, Waheed A. Al-masry, Fabrication and characterization of zinc oxide nanofibers for renewable energy applications, Volume 10, Supplement 1, February 2017, Pages S1067-S1072.

23. Alessandro Di Mauro , Massimo Zimbone , Maria Elena Fragalà , Giuliana Impellizzeri, Synthesis of ZnO nanofibers by the electrospinning process, Materials Science in Semiconductor Processing 42 (2016) 98–101.

24. Shaohong Wei, Meihua Zhoua, Weiping Du, Improved acetone sensing properties of ZnO hollow nanofibers by single capillary electrospinning, Sensors and Actuators B 160 (2011) 753–759.

25. Wei Wang, Huimin Huang, Zhenyu Li, Hongnan Zhang, Yu Wang, Wei Zheng, and Ce Wang, Zinc Oxide Nanofiber Gas Sensors Via Electrospinning, J. Am. Ceram. Soc., 91 [11] 3817–3819 (2008).

26. Seung-Hoon Choi, Guy Ankonina, Doo-Young Youn, Seong-Geun Oh, Jae-Min Hong, Avner Rothschild, and Il-Doo Kim, Hollow ZnO Nanofibers Fabricated Using Electrospun Polymer Templates and Their Electronic Transport Properties, acs nano, VOL. 3 ▪ NO. 9 ▪ 2623–2631 ▪ 2009.

27. Jin-Ah Park , Jaehyun Moon, Su-Jae Lee, Sang-Chul Lim, Taehyoung Zyung, Fabrication and characterization of ZnO nanofibers by electrospinning, Current Applied Physics 9 (2009) S210–S212.

28. Sujuan Wu, Qidong Tai, and Feng Yan, Hybrid Photovoltaic Devices Based on Poly (3-hexylthiophene) and Ordered Electrospun ZnO Nanofibers, J. Phys. Chem. C 2010, 114, 6197–6200.

29. Il-Doo Kim, Jae-Min Hong, Byong Hong Lee, and Dong Young Kim, Dye-sensitized solar cells using network structure of electrospun ZnO nanofiber mats, Applied Physics Letters 91, 163109 2007.

30. Yinhua Li, Jian Gong, Yulin Deng, Hierarchical structured ZnO nanorods on ZnO nanofibers and their photoresponse to UV and visible lights, Sensors and Actuators A 158 (2010) 176–182.

31. Zhengtao Zhu, Lifeng Zhang, Jane Y. Howe, Yiliang Liao, Jordan T. Speidel, Steve Smith and Hao Fong, Aligned electrospun ZnO nanofibers for simple and sensitive ultraviolet nanosensors, Chem. Commun., 2009, 2568-2570 | 2569.

32. Hui Wu, Dandan Lin, Rui Zhang, and Wei Pan, ZnO Nanofiber Field-Effect Transistor Assembled by Electrospinning, J. Am. Ceram. Soc., 91 [2] 656–659 (2008).

33. X. M. Sui, C. L. Shao, and Y. C. Liu, White-light emission of polyvinyl alcohol ZnO hybrid nanofibers prepared by electrospinning, Appl. Phys. Lett. 87, 113115 (2005).

34. Andrzej Stafiniak, Bogusław Boratynski, Anna Baranowska-Korczyc, Adam Szyszka, Maria Ramiaczek-Krasowsk, Joanna Prazmowska, Krzysztof Fronc , Danek Elbaum, Regina Paszkiewicz, Marek Tłaczała, A novel electrospun ZnO nanofibers biosensor fabrication, Sensors and Actuators B 160 (2011) 1413–1418.

35. Chang-Gyu Lee, Kyeong-Han Na , Wan-Tae Kim, Dong-Cheol Park, Wan-Hee Yang and Won-Youl Choi, TiO_2/ZnO Nanofibers Prepared by Electrospinning and Their Photocatalytic Degradation of Methylene Blue Compared with TiO_2 Nanofibers, Appl. Sci. 2019, 9, 3404; doi:10.3390/app9163404.

36. Ruilai Liua, Huiyan Ye, Xiaopeng Xiong, Haiqing Liu, Fabrication of TiO_2/ZnO composite nanofibers by electrospinning and their photocatalytic property, Materials Chemistry and Physics 121 (2010) 432–439.

37. Sun Hye Hwang, Jooyoung Song, Yujung Jung, O. Young Kweon, Hee Song and Jyongsik Jang, Electrospun ZnO/TiO$_2$ composite nanofibers as a bactericidal agent,Chem. Commun., 2011, 47, 9164-9166.

38. Ying Wang, Wenzhao Jia, Timothy Strout, Ashely Schempf, Heng Zhang, Baikun Li, Junhong Cui, Yu Leia, Ammonia Gas Sensor Using Polypyrrole-Coated TiO$_2$/ZnO Nanofibers, Electroanalysis 2009, 21, No. 12, 1432 - 1438.

39. M. E. Fragala,I. Cacciotti, Y. Aleeva, R. Lo Nigro, A. Bianco, G. Malandrino, C. Spinella, G. Pezzottid and G. Gusmano, Core-shell Zn-doped TiO$_2$-ZnO nanofibers fabricated via a combination of electrospinning and metal-organic chemical vapour deposition, CrystEngComm, 2010, 12, 3858-3865.

40. Jae Young Park, Sun-Woo Choi, Jun-Won Lee, Chongmu Lee, and Sang Sub Kim, Synthesis and Gas Sensing Properties of TiO$_2$-ZnO Core-Shell Nanofibers, J. Am. Ceram. Soc., 92 [11] 2551-2554 (2009).

41. Pingfan Du, Lixin Songa, Jie Xiong, Ni Li, Zhenqiang Xi , Longcheng Wang , Dalai Jin, Shaoyi Guo , Yongfeng Yuan, Coaxial electrospun TiO$_2$/ZnO core-sheath nanofibers film: Novel structure for photoanode of dye-sensitized solar cells, Electrochimica Acta 78 (2012) 392-397.

42. YUE Xue-Jun, HONG Tian-Sheng, XU Xing, LI Zhen, High-Performance Humidity Sensors Based on Double-Layer ZnO-TiO$_2$ Nanofibers via Electrospinning, CHIN. PHYS. LETT. Vol. 28, No. 9 (2011) 090701.

43. Hai Ying Wang, Yang Yang Xiang Li , Li Juan Li , Ce Wang, Preparation and characterization of porous TiO$_2$/ZnO composite nanofibers via electrospinning, Chinese Chemical Letters 21 (2010) 1119-1123.

44. Carina Chun Pei, Wallace Woon-Fong Leung, Enhanced photocatalytic activity of electrospun TiO$_2$/ZnO nanofibers with optimal anatase/rutile ratio, Catalysis Communications 37 (2013) 100–104.

45. Zheng Lou, Jianan Deng, Lili Wang, Rui Wang, Teng Fei and Tong Zhang, A class of hierarchical nanostructures: ZnO surfacefunctionalized TiO$_2$ with enhanced sensing properties, RSC Adv., 2013, 3, 3131–3136.

46. Muzafar A. Kanjwal, Nasser A. M. Barakat, Faheem A. Sheikh, Soo Jin Park and Hak Yong Kim, Photocatalytic Activity of ZnO-TiO$_2$ Hierarchical Nanostructure Prepared by Combined Electrospinning and Hydrothermal Techniques Macromolecular Research, Vol. 18, No. 3, pp 233-240 (2010).

47. Carina Chun Pei and Wallace Woon-Fong Leung, Enhanced photocatalytic activity of electrospun TiO$_2$/ZnO nanofibers with optimal anatase/rutile ratio, Catalysis Communications 37 (2013) 100–104.

48. Adem Yar, Bircan Haspulat, Tugay U″stu″n, Volkan Eskizeybek, Ahmet Avcı, Handan Kamı,s and Slimane Achour, Electrospun TiO$_2$/ZnO/ PAN hybrid nanofiber membranes with efficient photocatalytic activity, RSC Adv., 7, 29806–29814 (2017,).

49. Jian Li, Long Yan, Yufei Wang, Yuhong Kang, Chao Wang and Shaobo Yang, Fabrication of TiO$_2$/ZnO composite nanofibers with enhanced photocatalytic activity, J Mater Sci: Mater Electron (2016).

50. Evando S. Araújo, Bruna P. da Costa, Raquel A.P. Oliveira, Juliano Libardi, Pedro M. Faia and Helinando P. de Oliveira, TiO$_2$/ZnO hierarchical heteronanostructures: Synthesis, characterization and application as photocatalysts, Journal of Environmental Chemical Engineering 4 (2016) 2820–2829.

51. A.F. Lotus , S.N. Tacastacas, M.J. Pinti, L.A. Britton, N. Stojilovic , R.D. Ramsier and G.G. Chase, Fabrication and characterization of TiO_2–ZnO composite nanofibers, Physica E 43 (2011) 857–861.

52. Xian Liu, Yan-yu Hu, Ri-Yao Chen, Zhen Chen, and Hong-Chun Han, Coaxial Nanofibers of ZnO-TiO_2 Heterojunction With High Photocatalytic Activity by Electrospinning Technique, Synthesis and Reactivity in Inorganic, Metal-Organic, and Nano-Metal Chemistry, 44:449–453, 2014.

53. Delong Li, Xudong Jiang, Yupeng Zhang, and Bin Zhang, A novel route to ZnO/TiO_2 heterojunction composite fibers, J. Mater. Res., Vol. 28, No. 3, Feb 14, 2013.

54. Sun Hye Hwang, Jooyoung Song, Yujung Jung, O. Young Kweon, Hee Song and Jyongsik Jang, Electrospun ZnO/TiO_2 composite nanofibers as a bactericidal agent, Chem. Commun., 2011, 47, 9164–9166.

55. Ruilai Liu, Huiyan Ye , Xiaopeng Xiong and Haiqing Liu, Fabrication of TiO_2/ZnO composite nanofibers by electrospinning and their photocatalytic property, Materials Chemistry and Physics 121 (2010) 432-439.

56. Jin-Ah Park, Jaehyun Moon, Su-Jae Lee, Sang-Chul Lim and Taehyoung Zyung, Fabrication and characterization of ZnO nanofibers by electrospinning, Current Applied Physics 9 (2009) S210–S212.

Chapter-3
Fabrication and Characterization Techniques of Nanofibers

The advancement of science and technology has substituted conventional and cumbersome techniques of nanofibers fabrication that yields more accurate and reproducible outcomes. A valuable insight into different fabrication and characterization methods is essential in order to carry out effective work.

3.1 Fabrication Techniques of Nanofibers

In this section, we discuss about the various fabrication techniques which are being investigated and employed. These include drawing, template synthesis, phase separation, self- assembly, and electrospinning.

3.1.1 Template Synthesis

The template synthesis means that a template or mold is used to achieve a desired substance or form. A membrane of metal oxide having nano-scale pores is utilized in this synthesis process. This membrane is kept on a high viscous solution. This solution is perforated through the membrane upon the application of high water pressure. The polymer solution is brought into contact with the solution which transforms a polymer solution into nanofibers upon passing via the membrane. With this approach, the diameter of the nanofibers is the function of the pore diameter of membrane. **Figure 3.1** illustrates the steps involved in the template synthesis process.

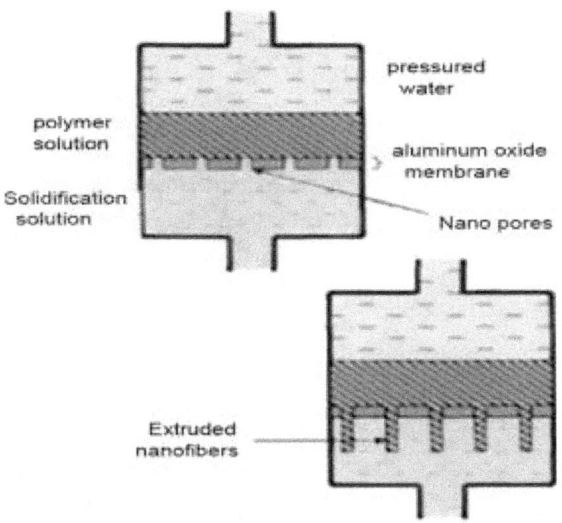

Figure 3.1. An illustration of template synthesis of nanofibers [1].

With this technique, nanofibers based on polymer, metal, semiconductors, or ceramics can be manufactured by using a nano-scale porous membrane. This technique can be processed either by the chemical or electrochemical oxidative polymerization process and unable to grow nanofibers of longer length. However, the diameter of these fibers is controlled by the used membrane's pore size [2,3].

3.1.2 Drawing

Drawing is another means of processing fibers. In a simple word, it is like dry spinning. This approach often includes a sharp tip/micropipette so it is advantageous. A sharp tip is used in this process to extract a droplet from an originally formed solution of polymer as a liquid fiber. The solvent is then evaporated because of the high surface area and enables the liquid fibers to solidify.

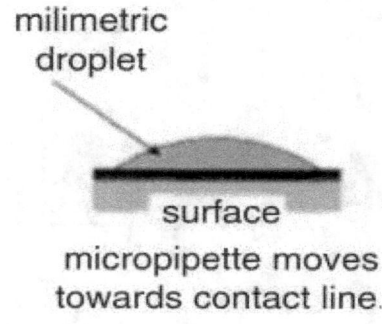

micropipette moves towards contact line.

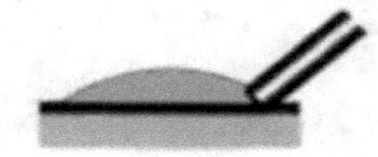

micropipette comes into contact with contact line.

withdrawal of micropipette produces nanofiber.

Figure 3.2.Step-by-step drawing process for the preparation of nanofibers.

Rather than using a sharp tip with a continuous dose of the polymer, the hollow glass micropipettes may be employed to prevent the volume shrinkage issue which restricts the continued fiber drawing and influences fibers diameter as shown in **figure 3.2** [4]. After the micropipette is submerged into the droplet via a micromanipulator, it is gently pulled out the liquid and driven at a low rate (about 10-4 m/s), thereby pulling, and depositing nanofibers on the surface by contacting it at the edge of the micropipette. On every droplet, this cycle is repeated number of times to prepare the nanofibers [5]. With this technique, continuous nanofibers can be prepared. In addition to the precise control of the main parameters the drawing speed and viscosity enables the reproducibility and control over the diameter and length of the produced fibers [6].

Even though this is a straight forward process, but is restricted to the laboratory level as the nanofibers prepared one by one. This is a discontinuous method with poor productivity means it supports the preparation of one by one single nanofibers. This technique has the ability to monitor the diameter and length of the prepared nanofibers. In this way, viscoelastic material alone can survive the pulling force, and only fibers greater than 100 nm in diameter can be generated according to the size of the orifice.

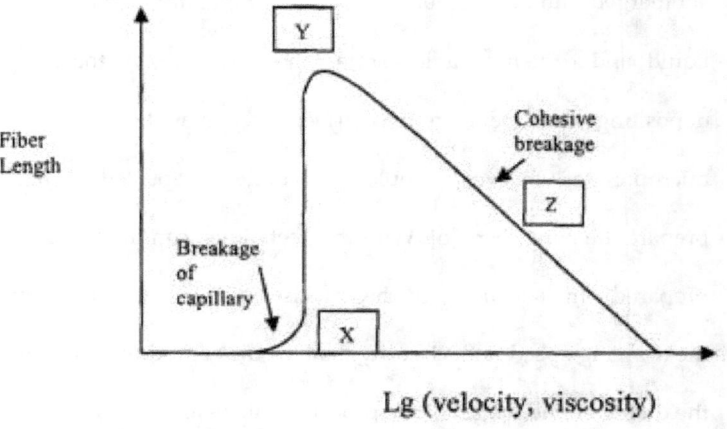

Figure 3.3.Step-by-step drawing process for the growth of nanofibers [5].

At the outset of evaporation, the drawn fibers may break due to the Rayleigh fluctuations and indicated with as stage 'X' in the curve as shown in **figure 3.3**. Nanofibers drawn during the second stage of evaporation is indicated as stage 'Y' in the curve. During the droplet's last evaporation phase as indicated as stage 'Z', the solution gets centralized on the droplet edge and split cohesively. The drawing of fibers thus includes a viscoelastic substance which can be deformed tightly and which is adequately compact to withstand the stresses during the pulling. The drawing method can be seen as dry molecular spinning [7].

3.1.3 Phase Separation

With this approach, phases are distinguished by the physical incompatibility. The solvent phase is derived from the solution whereas the other phase is still present.

There are four key steps of this approach:

1) Dissolution of polymer in solvent either at room temperature or at higher temperature.

2) Gelation which seems to be the hardest stage to monitor the morphology of nanofibers.

3) Extracting solvent from the gel using water.

4) Freezing and freeze drying in a controlled vacuum environment.

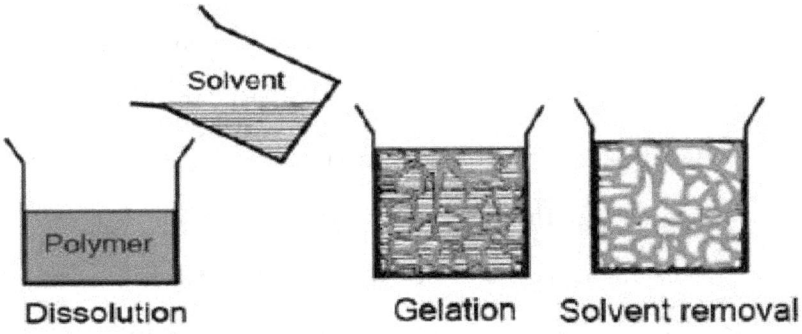

Figure 3.4. Phase-separation approach for the preparation of nanofibers [8].

Polymer concentration influences the characteristics of nanofibers. An increased polymer concentration yielded the reduced porosity while enhanced mechanical strength of the fibers [3]. Initially, this approach is focused on the preparation of the homogeneous solutions by diluting the polymer at ambient environment and then producing gel preserving the solution's gelation temperature in which nanofibrous matrices are formed through phase separation, and removing the solvent and freeze-drying the matrix contributing to nanofiber forming as illustrated in **figure 3.4**.

It is a basic production infrastructure necessity. A nanofibers matrix can be produced directly in which the mechanical properties of this matrix can be modified by adjusting the concentration of polymers [5]. To date, a limited fabrication of nanofibers is being reported using the phase separation process. This technique is applied by using few polymers such as PLA and PGA [7]. This process does not create long continuous fibers with all the polymers. Since it has to be gelated so this restricts the use of phase separation technique.

By using this technique, preparation of nanofibrous poly(L-Lactic) acid (Pl-LA) has been reported. They have followed the following five steps: dissolution of polymer, b) gelation, c) extraction of solvent, d) freezing and (e) freeze drying. The freeze drying process is explored in following steps. Tetrahydrofuran has been added to poly (L-lactic) acid to make the solution by preferring the predefined concentration. Over two hours, the solvent was stirred at the temperature 60 ºC in order to get the homogenous solution. Two milliliters of the solution is placed in a Teflon vial at SCC and kept in the refrigerator while maintaining the temperature of gelation from -18 ºC to 45 ºC). The prepared gel was held for two hours at gelation temperature. The gel-containing vial was soaked in pure water for solvent exchange and the solution while water was several times changed. The gel was then drained, covered in filter paper, and then kept in a freezer at -18 ºC for two hours. Ultimately, the gel was transferred into a frost-drying flask and at temperature -55 ºC freeze-dried under vacuum.

3.1.4 Self-Assembly

The process of self-assembly is also known as the bottom-up fabrication approach by means of molecules are assembled and formed into shapes or frameworks by non-covalent forces such as hydrogen bonding, hydrophobic forces and electrostatic reactions [4]. It's a useful approach for preparing thin (<100 nm to few nm) along with several nm long nanofibers by the mechanism of self-assembly process of small molecules by poor interactions including hydrogen bonding and hydrophobic interactions.

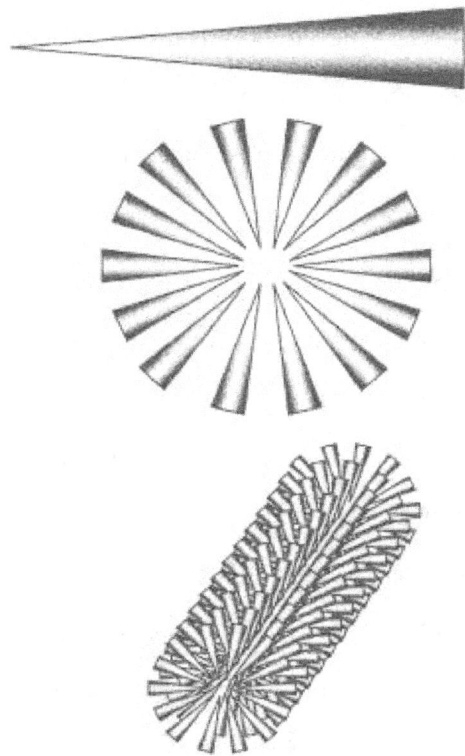

Figure 3.5. An illustration of self assembly of preparation of nanofibers [10].

The key process is oriented to the intermolecular forces which bind together small units of molecules; the configuration of small molecular units defines the macromolecular nanofibre's over form. The main drawback of the system is that it is a complicated, lengthy and highly intricate process with the lower productivity and a lack of the uniform growth of the nanofibers Therefore, it is a process that is restricted to the processing, by itself or under an external stimulus, of nanofibers from small active molecules [7,9].

Figure 3.5 depicts the schematic on self-assembly process for the preparation of nanofibers. In this case, a small molecular (refer top figure) is formed in a concentric manner, so that association may shape arming of concentratedly organized tiny molecules (refer middle figure). This expansion in the normal plane provides the nanotomic axis (refer bottom figure), which is a main mechanism for a typical self-assembly.

3.1.5 Electrospinning

The most effective and easy method for the processing of ultra-thin nanofibers is electrospinning. This technique uses a high voltage source. The metal collector has grounded it to the negative end of the voltage supply, the positive end of the connection is associated with the wire. It produces the potential difference between the two sides, speeding up the polymer fluid in the shape of a jet solution from the needle to the collector. The solvent gets evaporated prior to reaching the collector surface and stored as an interconnected web of the fibers. Due to its surface tension, which creates a voltage on the solvent layer, the polymer fluid in the needle is held up. The repellence and contraction of the voltage between the surface charge and its counter electrode corresponds to a force directly opposite the surface tension. At the end of the needle, it forms a hemispherical surface of solution which extends to form a conical form called „Taylor cone „particularly when the applied voltage is exceeded. When the electrical energy is further increased, the repulsive electrostatic power exceeds the solution's surface tension and the jet is throughout from the Taylor cone. The ejected polymer solution undergoes a cycle of instabilization and elongation that renders the jet very long and small. Meanwhile, the solvent is evaporated and leaves a charged polymer fiber. Different processing parameters influence the diameter of nanofibers. Such criteria include voltage, tip to collector distance feeding intensity and concentration. Thanks to its flexibility, cost-effectiveness and the regulation of the multiple processing parameters it is a benefit over other methods; the nanofibers dimensions can be monitored.

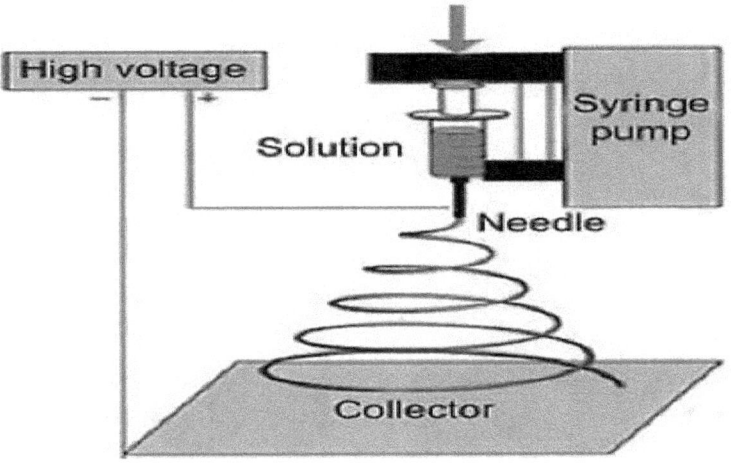

Figure 3.6.A schematic of electrospinning setup for the preparation of nanofibers.

A schematic view of traditional electrospinning set-up is shown in **figure 3.6**. A standard electrospinning system is composed of three major components: a high-voltage supply which charges the polymer solution, a syringe with pumps which controls the flow rate and the polymer solution is supplied via a capillary link and the collector is grounded onto which nanofibers are collected. A syringe pump with a steady level is used in this method to draw the solution (solution, polymer or melted fluid) from the tip of the needle.

In summary, capabilities, merits, and demerits of various preparation techniques of nanofibers are briefly discussed in **Table 3.1** and **Table 3.2**.

Table 3.1 Capabilities of various nanofibers preparation methods.

Technique	Fiber scalability?	Reproducibility?	Process Convenient?	Diameter Tunability?	Technical Remark
Template Synthesis	No	Yes	Yes	Yes	Limited to Lab/Research
Drawing	No	Yes	Yes	No	Limited to Lab/Research
Phase Separation	No	Yes	Yes	No	Limited to Lab/Research
Self-assembly	No	Yes	No	No	Limited to Lab/Research
Electrospinning	Yes	Yes	Yes	Yes	Potential demand in industries.

72

Table 3.2 Merits and demerits of various nanofibers preparation methods.

Technique	Merits	Demerits
Template Synthesis	Fiber having different diameters can be quickly prepared	-
Drawing	Easier method than other and simple equipment	Yield discontinuous nanofibers
Phase Separation	Simple equipment, easy preparation of nanofibers matrix, reproducibility, strength of the nanofibers can be tuned via polymer concentration	Compatible with selected polymers
Self-assembly	Fit for nanofibers of smaller length	Process is completive
Electrospinning	Inexpensive, long and continuous production of nanofibers, reproducibility etc.	Jet instability issue

Certain points should be considered for the successful preparation of nanofibers using electrospinning. The choice of suitable solvent is important in order to fully dissolve the polymer. The solvent vapor pressure must be sufficient so that the fibers evaporate rapidly enough to preserve their integrity if the fibers reach to the collector but not so fast to enable fibers to roughen before entering the nanometer scale. The solvent's surface viscosity and friction must be maintained low to keep the continuous flow of jet or minimal enough to enable the solution to drain out of the pipette. The strength is needed to be sufficient to exceed the polymer solution's viscosity and surface tension and to shape and support the jet from the pipette. The distance between the pipette and the grounded target is needed enough so the sparks generation can be avoided but must be adequate to evaporate the solvent in time to make the fibers.

3.2 Characterization Techniques

The proper familiarity of analytical tools is essential to ensure the efficient use of the equipment. In this view, this part describes the various analytical techniques used for the investigation of nanomaterials.

3.2.1 X-Ray Diffraction (XRD)

X-ray powder diffraction (XRD) is a fast and non-destructive technique that is basically employed for the detection of the crystalline phase present in the sample and for the collection of details of unit cell dimensions [11]. Initial investigation boosted the discovery of XRD after reporting the formation of crystalline planes in a crystal mesh as 3D diffraction gratings after illuminating X-rays. X-rays diffraction system is a widely used tool for the investigation of atomic spacing and crystal structures.

Working Principle and Instrumentation

XRD technique works on the principle of constructive interference caused by the interaction between monochromatic rays and the crystalline material. Such x-rays are emitted by a cathode ray tube, filtered to create monochromatic light, collimated to focus, and further directed to the specimen. When conditions meet the Bragg Law, the interaction of the incident rays with the specimen causes constructive interference [12,13].

Bragg law is represented as

$$n\lambda = 2d \sin \theta$$

where *n* is an integer which represents the order of the diffracted beam, λ is the wavelength of the x-ray beam, *d* is the distance between the neighboring planes of the atom and θ is the incident angle. The d-spacing can be determined by using the obtained θ whereas λ is constant. The distinctive collection of d-spacing and their intensity produced in a typical x-ray scan provide a specific "fingerprint" of the phases existing in the specimen. **Figure 3.7** depicts the principle of x-ray diffraction.

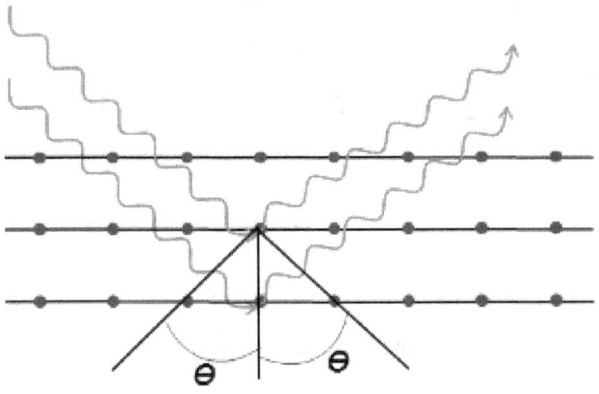

Figure 3.7 An illustration of X - Ray Diffraction.

The Bragg's law refers to the wavelength of light rays radiation to the angle of the diffracted angle and the lattice spacing arranged in the crystalline specimen. Such x-rays are detected, monitored, and recorded. Through scanning the sample in a range of 2θ angles, the lattice diffraction from all directions of the powder sample. The transformation of the diffraction peaks into d-spacing helps the mineral to be classified since individual mineral has a specific d-spacing framework. This is generally done by matching d-spacing with standard reference trends. An X-ray tube is a source of all diffraction techniques. Such X-rays are aimed at the specimen whereas the diffracted rays are gathered [11-13].

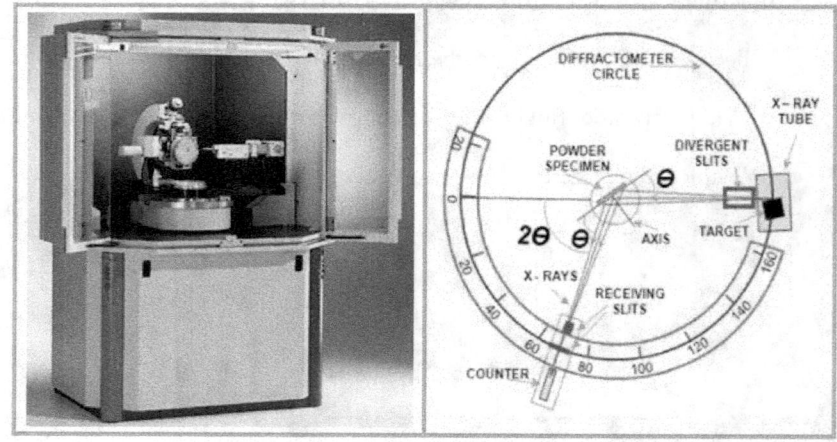

Figure 3.8. XRD unit (left) and X-Ray Diffractometer (right) [13].

There are three elementary components of x-ray diffractometer: x-ray tube, sample holder, and x-ray detector. A picture of Bruker's D8-Discover X-ray diffraction instrument is shown in **figure 3.8**. X-rays are created upon heating a filament located in the cathode ray tube which is used to generate the electrons. These electrons are accelerated towards the specimen with the application of a voltage as a result the specimen is overwhelmed with these electrons. If electrons have enough energy to dislocate internal shell electrons from the specimen material, a characteristic x-ray spectrum is generated. In order to produce monochromatic x-rays needed for diffraction, filtering by foils or crystal monochromters is essential. The wavelength of CuKα radiation is 1.5418Å while copper is the typical target source for the single-crystal diffraction. The collimated X-rays guided towards the specimen. The reflected x-rays intensity is collected when the specimen and detector are rotating. Whenever the geometry of the X-rays focusing on the sample is fulfilled the Braggs condition, then the constructive interference causes the generation of intense peaks. A detector captures and analyses this X-ray signal and transforms it to a count rate. The X-ray diffractometer's mechanism is to move specimen at an angle θ in the direction of the collimated beam. The detector is placed on an arm to capture the diffracted x-ray and moves at an angle of 2θ. The tool used to sustain the angle and rotate the specimen is called a goniometer.

3.2.2 Ultra-Violet Visible Spectroscopy (UV-vis)

An absorption spectroscopy measurement uses electromagnetic radiation from the range from 190-800 nm [14-19]. As ultraviolet or visible light absorption by a molecule gives rise to the transition between the molecule's electronic energy levels therefore, it is also referred as electronic spectroscopy.

Molecule's total energy is the sum of three components that is electronic, vibrational, and rotational energies. After absorbing the UV light, the energy levels of molecules get alters. The molecules containing π-electrons or free electrons absorb UV-vis light and excited the electrons. The energy transformations between bonding states and anti-bonding electronic states are depicted in **figure 3.9**.

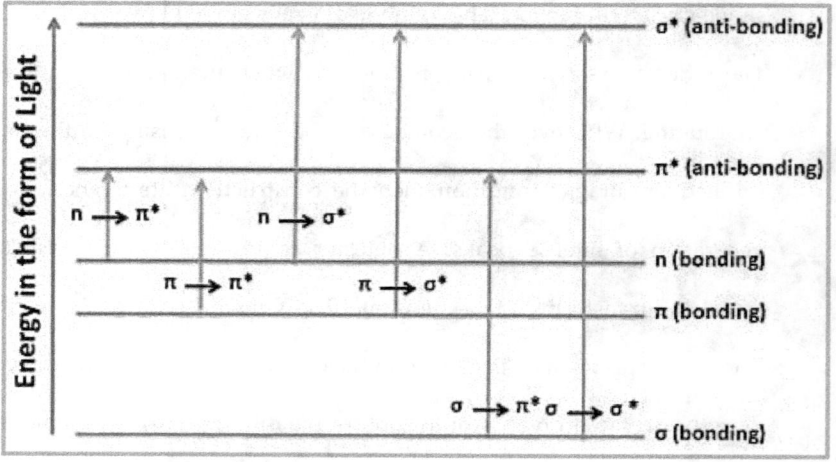

Figure 3.9. An illustration of energy level transitions between electronic states in a molecule.

The Beer –Lambert law is the heart of the absorbance spectroscopy and the equation is represented by

$$A = \varepsilon b c$$

Where ε is the molar absorptivity, A is absorbance (arb. units), b is the cuvette width, and c is the solutions concentration.

Working Principle and Instrumentation

UV-vis spectrometer contains five elements, including source, monochromatic, sample holder, detector, and the single processing unit. A simple mechanism of UV-vis setup is depicted in **figure 3.10**. This technique is used for the quantitative measurement of the material's absorption/transmission/reflection under the test by varying the wavelength from 190-110 nm [14]. The schematism of dual beam UV-vis spectrometer is depicted in **figure 3.11**.

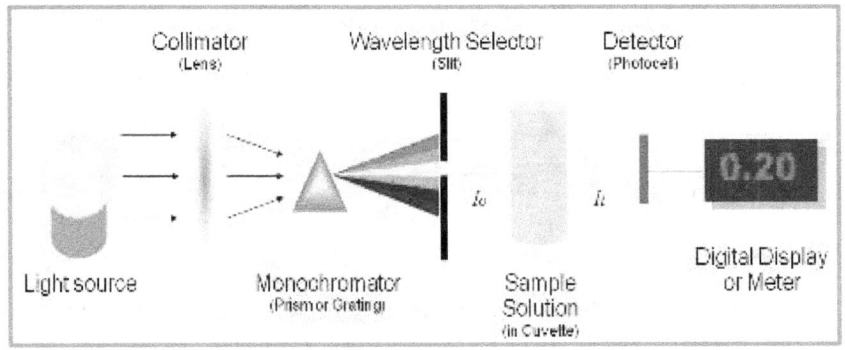

Figure. 3.10. Optics of UV-vis spectrophotometer.

Basically, UV-vis machine is equipped with two light sources that are deuterium discharge lamp for UV range while halogen lamp for visible and near-infrared range measurements. During scanning, these two lamps get auto swap for the UV to the visible range. Once testing is started, the UV-vis light generated by the respective sources passes through the slits and inserted in the monochromator.

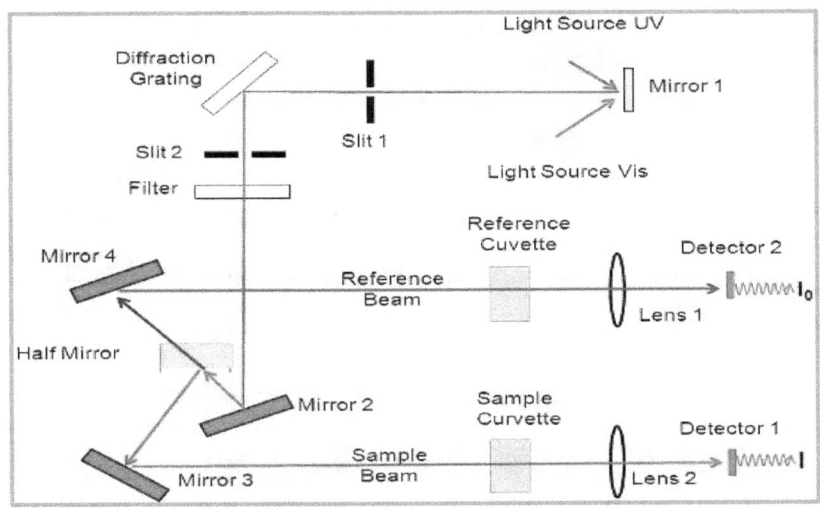

Figure 3.11. Schematic diagram of UV-vis spectroscopy.

The beam collimation takes place in order to target the sample at an angle. The signal is separated by the grating or prism into its wavelengths. Through rotating the dispersing element or the output slit, radiation of a single wavelength is permitted to leave through the output slit. This light beam is divided into two halves and after crossing through a series of mirrors. One part of the beam crosses through the specimen while another part goes through the reference. Both the specimen and reference are loaded in transparent cuvettes. These cuvettes are polished ones in order to reduce the losses which may occur due to reflection and scattering during measurement. The respective beams passing through the specimen and reference directed towards the detector. A widely used detector is the photomultiplier tube, which deals or intensifies the resultant signals. Finally, the detector is attached to a computer to visualize and printer to print the obtained result.

3.2.3 Fourier Transform Infrared Spectroscopy (FTIR)

Fourier Transform Infrared Spectrometer (FTIR) is a spectral instrument used to gather and digitizing the interferogram, executing the FT function, and showing the spectrum [14,15,17]. The FTIR basically illuminates the sample and measures the reflected, transmitted, or absorbed light for a wavelength range. The FTIR analysis applies to quantitative as well as qualitative applications. It can be used to measure solids, liquids, and gasses. The existence of functional groups in a molecule is easily identifiable using FTIR. In fact, the exclusive set of absorption peaks can be used to reveal the identity of a single substance or to identify different impurities. A digital image of the FTIR equipment is shown in **figure 3.12**.

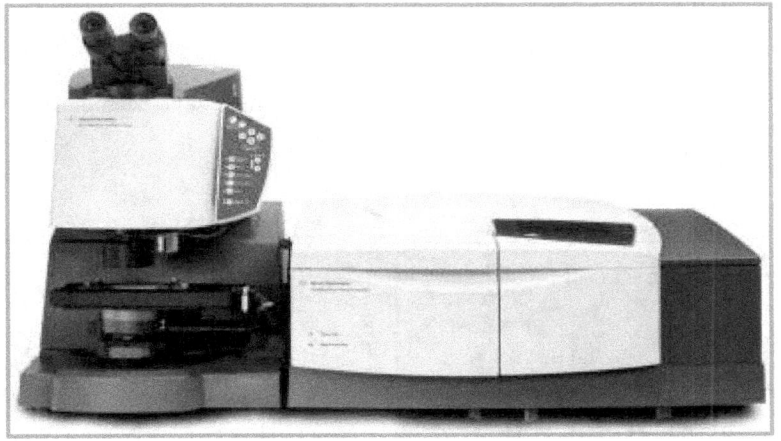

Figure 3.12. Fourier Transform Infrared Spectrometer unit.

Working Principle and Instrumentation

The IR theory lies in the fact that chemical bonds in a molecule are excited in a broad frequency spectrum. This happens when a molecule subjected to IR radiation absorbs IR which is representative of the molecule. The charge disparity in the electric fields of the atoms in a molecule induces a total dipole moment in the molecule. The total dipole moment molecules cause the interaction of infrared photons with the molecule inducing excitement at higher vibrational conditions. When the sample exited by IR radiation, the transmission, and reflection of the IR radiation at various frequencies is transformed into an absorption spectrum.

A typical FTIR system has three essential components which include an IR source used for radiation, interferometer, and detector. **Figure 3.13** depicts a simple optical model of FTIR spectrometer. The IR source based on an inert solid excited from the temperature range 1000-1800 ° C. There are three light sources are being investigated which have been employed in FTIR [18-22].

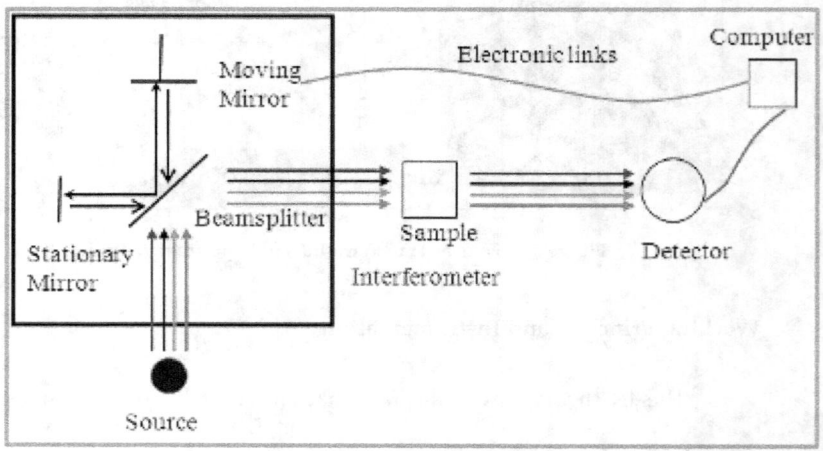

Figure 3.13. An illustration of FTIR Spectrophotometer.

These sources yields constant radiation however, their energy profiles are distinct. The interferometer separates radiant rays, produces an optical path difference between rays, and regroups them to create repeated interference signals, determined by a sensor with respect to the optical path distance. The interferometer causes interference signals by adjusting the relative position of the moving mirror to the fixed mirror, which produces an interference pattern. The resultant beam hits the specimen and then directed towards the detector. For the FTIR analysis, potassium bromide (KBr) pellet approach is employed to prepare the specimen. As KBr is transparent for the IR light therefore, it is usually used. The specimen is milled into fine powder using KBr. The KBr is known to be extremely hygroscopic therefore, rapid preparation of pellet is recommended. This material is then pressed to get a thin pellet which is loaded in the FTIR for the testing. The specimen concentration is limited in KBr from the range 0.2-1.0 %.

The interferometer is the backbone of the FTIR spectrometer. **Figure 3.14** shows the Michelson Interferometer system used in FTIR.

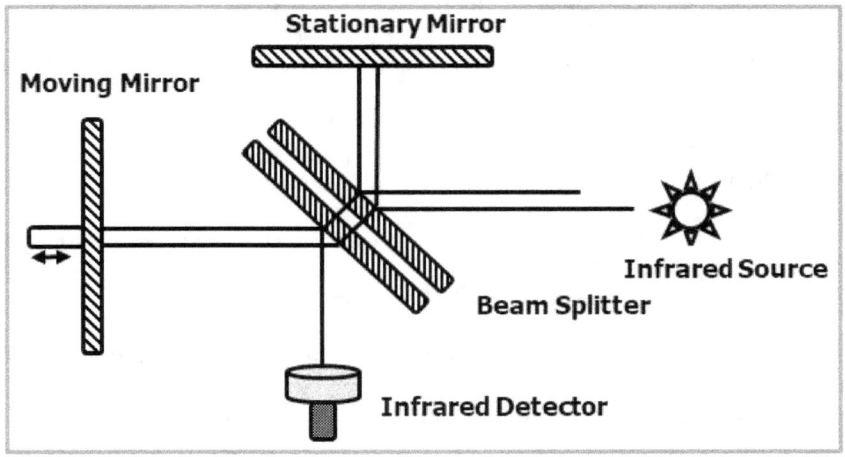

Figure 3.14. Schematic of Michelson Interferometer employed in FTIR Spectroscopy.

This includes an IR lamp, detector, beam splitter, and two mirrors. The beam splitter is the core of interferometer and is basically a partial silvered mirror. The splitter reflects one portion of irradiated light while concurrently transmits remaining light. This split light beam travels through a moving mirror. The stationary and mobile mirror reflects all light beams back to the beam splitter, which again partially reflects and transmits each beam. In this way, one part of beam goes to detector whereas another one towards the source.

3.2.4 Raman Spectroscopy (RS)

Raman spectroscopy is well recognized approach to investigate the large number of matters such as solids, liquids, and gases. Raman spectroscopy technique comes under the non-destructive methods in which sample treatment is not required [15,17]. A typical instrument of Raman spectroscopy is depicted in **figure3.15**. FTIR is the most common method used for the detection of unknown organic materials. In a similar way, Raman spectroscopy is also employed to collect the molecular bonding details. But Raman study is more useful when IR analysis is limiting the facts of the sample. Both techniques are similar and can provide accurate data of unknown materials when these are connected together. Further, a combination of Raman spectroscopy with a confocal microscope can yield depth profile analysis which is useful to collect details of the material beneath the surface.

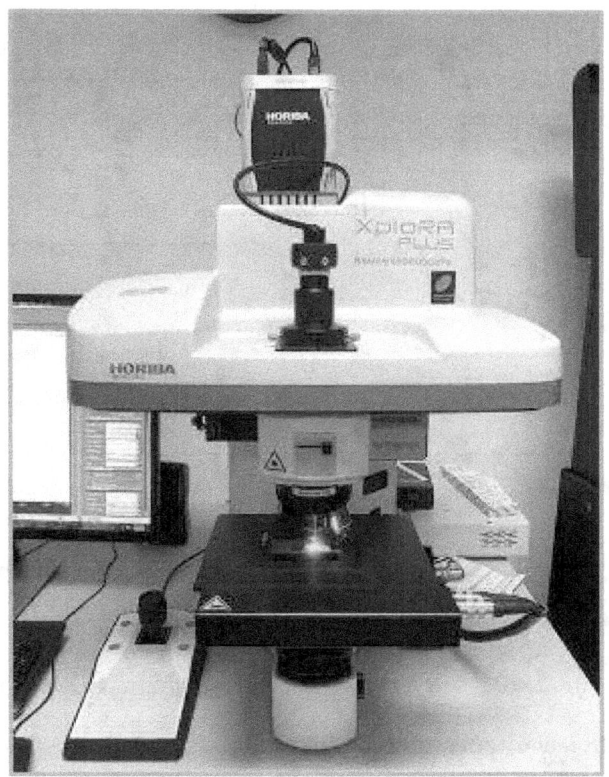

Figure 3.15. Raman Spectroscopy unit.

This technique involves the illumination of a sample with monochromatic light and the scattered light is examined using a spectrometer[18].

Working Principle and Instrumentation

It mainly consists of a laser, microscope, detector, beam splitter, and notch filter. The laser is used as the source of light and the notch filter is used to transmit the beam from a very narrow range based on the visible frequency of the laser radiation.

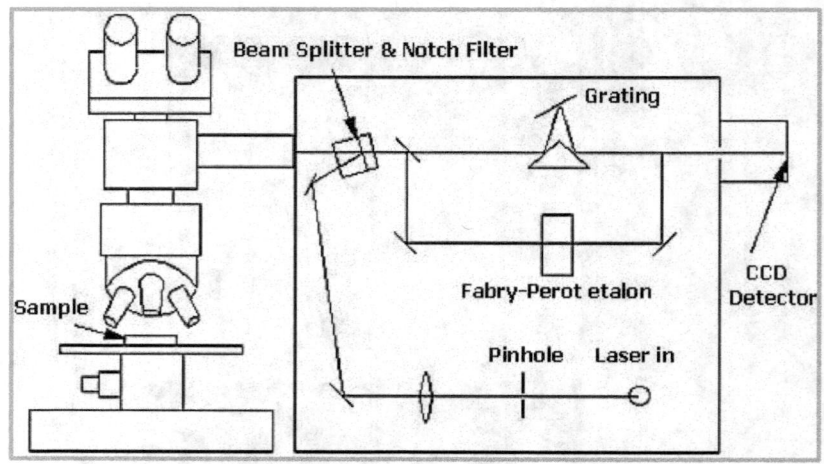

Figure 3.16.Schematic of Raman Spectrometer.

Figure 3.16 shows the schematic of the Raman spectrometer. It is attached with a microscope which is useful define the area for the analysis. The irradiated light comes from the specimen which goes into the spectrometer after passing from the optics. The attached CCD detector captures the Raman shift which is then processed by the computer and the graph is plotted [22]. During the testing, either elastically or in elastically scattering happens while later one is mainly contribute for the Raman study. The obtained graph of photon number versus Raman shift is known as Raman spectrum.

3.2.5 Thermogravimetry/Differential Thermal Analysis (TG/DTA)

Thermal activity is one of the physical characteristics of any substance which allows the analysis of a component in terms of its quality and application of the substance. TG/DTA is known as the concurrent thermal analyzer which is capable of characterizing several thermal properties in a single run. TG part reveals the various temperature zones of decomposition, reduction, and oxidation states of the specimen under the test. This also tests weight fluctuations correlated with decomposition and oxidation along with physical or chemical modifications resulting in weight loss or gain of the sample. The DTA part indicates the occurrence of decomposition is either by the endothermic or exothermic. DTA also tests temperatures leading to phase shifts where there is no mass loss including melt, crystallization and glass transformations [23,24].

The TG/DTA measurement is performed under the controlled environment of gas and the temperature. It performs the weight loss of the specimen with respect to the temperature and time. The weight change by percentage across a temperature range allows physical or chemical mechanisms to be observed that have induced the specimen's weight loss or gain [25].

The DTA metrics the useful information of specimen temperature (T_s) and the reference temperature (T_r). The T_s-T_r graph shows a sequence of peaks or phase shifts over specified temperature ranges, which track the temperatures during which thermal mechanism takes place [26].

Working Principle and Instrumentation

As described above, TG is a mass-change measurement technique that relates the decomposition, oxidation, and evaporation of the specimen in accordance the rise in temperature and time.

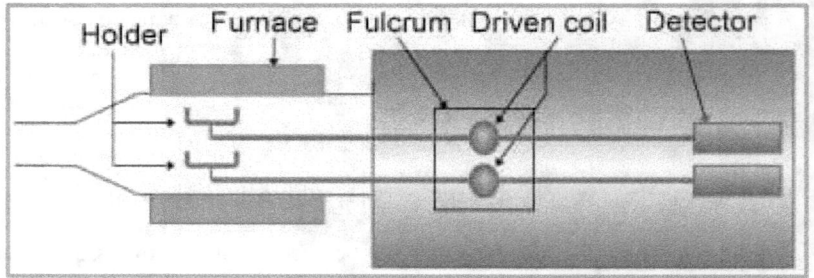

Figure 3.17. Schematic of TG/DTA system.

Referring to **figure 3.17**, there are two pans can be found in the furnace which is used to hold the specimen and reference. These pans are connected to the sensitivity-calibrated drive coils which are used to determine the masses of the specimen and reference. The variation in mass is measured through TG signal. The consequences of beam extension and convection flow are balanced with the produced differential mass. It makes extremely sensitive TG measurement. The measure of specimen mass and reference via the standalone drive coils facilitates the correction of the TG baseline.

In addition, each holder is fitted with a thermocouple that facilitates concurrent output of DTA signal. To investigate the thermal decomposition, oxidation, dehydration, heat resistance, and kinetic study TG is effective. Different data can be obtained from one specimen by integrating this with the additional technique. Simultaneous measuring TG/DTA system is extremely common [27].

3.2.6 Field-Emission Scanning Electron Microscope (FESEM)

Field-Emission Scanning Electron Microscope (FESEM) is a microscope that uses electrons i.e. particles with negative charges rather than light. The electrons are emitted through a field emission source while the target is processed in a zig-zag fashion by electrons [28]. A FESEM is used to visualize the surface or whole or fractional objects with very small topographical details. This method is used by experts in genetics, chemistry, and physics to analyze structures that can be tiny or upto 1 nm. FESEM is exploited to analyze organels and DNA content, synthetic materials, coatings etc. for their morphological investigations.

In contrast with SEM, FESEM offers a wide range of sample surface's detail, but with a higher resolution. The key difference among FESEM and SEM is the method of electron generation The FESEM uses a field emission gun that delivers high-and lower-energy electron beams that boost spatial resolution dramatically and make operation at extremely low potential (0,02 to 5 kV). It helps to reduce the charging effects on non-conductive materials and avoid damage to samples prone to electron beams. FESEM using in-lens detectors is another extremely extraordinary feature. Such detectors, which are configured for high-resolution function and very low acceleration potential, are important to achieve full equipment performance [29-32].

Working Principle and Instrumentation

In FESEM, the electrons emitting from field emission source gets stimulated in a large electrical field gradient. In turn, electrons are directed and deflected through electronic lenses into the high vacuum column to build a narrow scan stream, which targets the sample. As a consequence, secondary electrons are released from every location of the sample. The angle and speed of these released electrons are related to the sample's surface. The detector detects these electrons and generates signal which intensified and digitalize in computer screen.

The typical FESEM system is depicted in **figure 3.18**. It has a cylindrical column ,marked as 1' which helps to focus the electrons. There is a column knob provided which regulates the electron beam step by step. Few tubes are attached in the instrument and cryo-unit to maintain the vacuum and temperature. The microscope works via steering „marked as 2'. A cryo unit „marked as 3' is positioned left of the column and has a binocular (3). When the exchange chamber is placed in front of typical (not cryo) microscopy, the sample is inserted into the high vacuum region below columns (4).

Figure 3.18. A typical FESEM system.

The target can be screened on the large screen „marked as 5' during the scanning. The small screen „marked as 6' is used to view the chamber under which the object is kept. The PC for archiving and editing images is on the right hand side „marked as 7'. The cabinets underneath the desk have electronic controls „marked as 8'.

Sample Preparation

First, gold sputtering of thickness from 1.5-3.0 nm is done to make the conductive specimen. The coating unit is shown in **figure 3.19**. To prevent disruption to the delicate structures due to surface tension, chemically attached contents must first be eliminated and dry out not over the critical temperature. In a separate unit, the coating is then done.

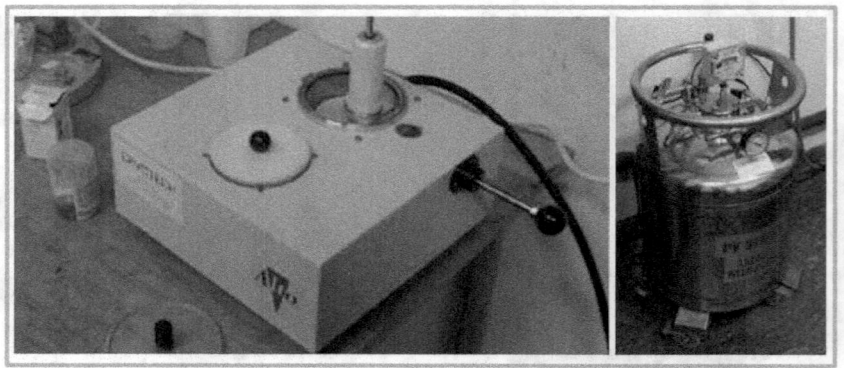

Figure 3.19. Coating unit used for FESEM sample preparation (left) and gas cylinder (right).

Electron Source

Electrons are produced mostly through heating a tungsten filament to the temperature 2800 °C by applying the current. A lantanumhexaboride (LaB) crystal, which is placed on a tungsten wire releases electrons. **Figure 3.20** shows the typical optical system of SEM and TEM.

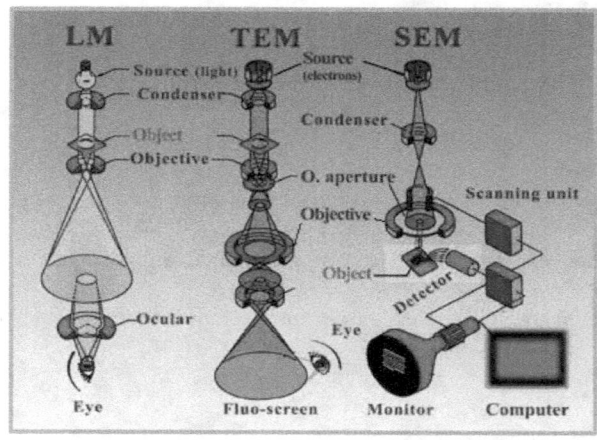

Figure 3.20. Schematic diagram of FESEM machine.

The only cold source is employed in the FESEM as compared to heating one in the conventional microscope. A sharp tungsten serves as cathode against the primary and secondary anodes while the cathode-anode voltage 0.5 to 30 kV is used.

3.2.7 Transmission Electron Microscope (TEM)

Transmission Electron Microscope (TEM) uses energetic electrons to deliver sample's morphological, compositional, and crystallographic details. TEM is the most effective microscope which provides magnification upto 1nm. TEM produces high-resolution images in two-dimensional that allow its broad range of applications in educational, science and industrial uses.

Working Principle and Instrumentation

TEM microscope creates a high-resolution picture when the energetic electrons are interacted with the sample under the vacuum condition. Air must be evacuated from the chamber to make room for transmission of electrons. Further, these electrons travel by electromagnetic lenses. These solenoid tubes are wrapped with a coil. The beam passes through the solenoids, descends the column, and contacts the monitor, where the electrons are transformed to light, which then generates a picture. The picture is corrected by controlling the gun voltage so that electrons velocity can be increased or decreased and electromagnetic wavelength can be changed by the solenoids [33,34].

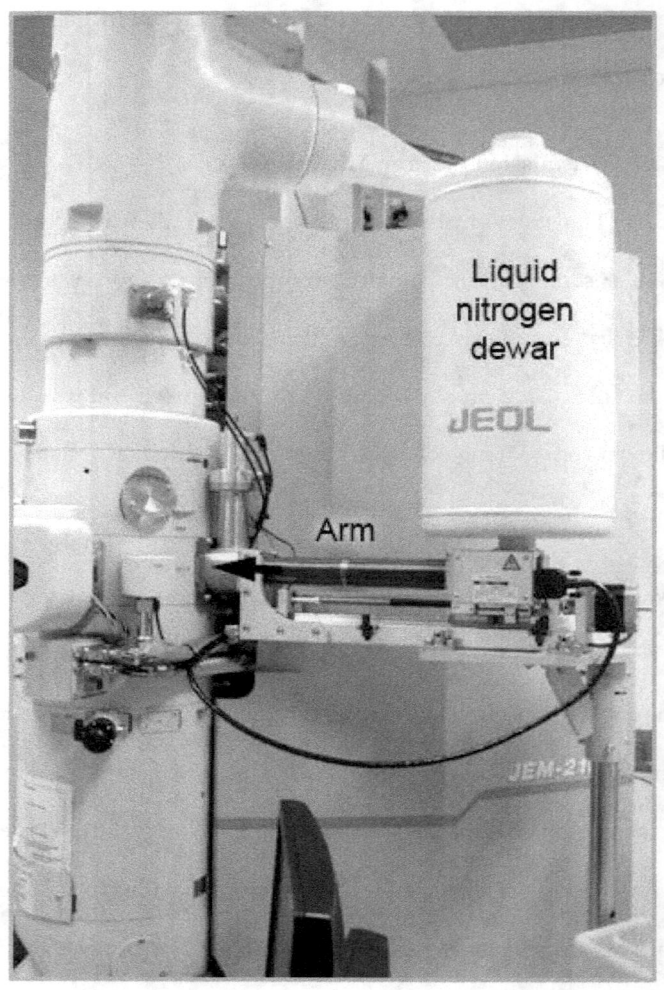

Figure 3.21. TEM detector unit.

The images are focused on a screen or photographic plate through the coils. The velocity of the electrons corresponds to the wavelength of the electrons during transmission. The stronger the electrons, the smaller the distance, and the superior picture clarity and information. The brighter areas of the picture indicate the positions in which more electrons may cross the sample, and the darker areas represent the dense areas of the specimen. It provides the shape, texture, type, and size of the sample with detail. Samples need certain properties for obtaining a TEM investigation. The specimen is needed to be diced very thin so that electrons can pass through it. Samples must withstand the vacuum chamber and need special preparation before being examined. The sample preparation involves the dehydration, non-conductive sputter coating, cryofixation, sectioning and staining methods [35- 37].

3.2.8 Electron Dispersive Analysis of X-rays (EDAX)

The concept driving EDAX is much similar to the microanalysis of electron probes. When an electron beam hits a sample, a proportion of electrons energize the sample's atoms which cause the emission of x-rays as a result of the transition to their ground level. The strength of these x-rays is directly associated with the atomic number of the excited elements, and thus their measurement is the foundation of elemental investigation [38].

In this system, the X-ray detector is usually a lithium-driven p-I-n silicon diode which is maintained at the liquid nitrogen temperature. The cooled detector is shielded by a beryllium window. Through multi-channel analyser the output pulses are processed. The spectrum study is conducted using the energy dispersion process. This approach is able to investigate the elements listed in the periodic table after Na. Sometimes 0.1-0.01 atomic percentage is a measurable electron limit for homogenously dispersed materials [39].

References

1. Ahmed Barhoum, Mikhael Bechelany and Abdel Salam Hamdy Makhlouf, Handbook of Nanofibers, Springer Nature Switzerland AG 2019
2. Zhang X, Lua Y (2014) Centrifugal spinning: an alternative approach to fabricate nanofibers at high speed and low cost. Polym Rev 54:677–701.
3. Kumar P (2012) Effect of collector on electrospinning to fabricate aligned nanofiber. Department of Biotechnology & Medical Engineering National Institute of Technology, Rourkela.
4. Beachley V, Wen X (2010) Polymer nanofibrous structures: fabrication, biofunctionalization and cell interactions. Prog Polym Sci J 35(7):868–892.
5. Ramakrishna S, Fujihara K, Teo W-E, Lim T-C, Ma Z (eds) (2005) An introduction to electrospinning and nanofibers. World Scientific Publishing, Singapore.
6. Nayak R, Padhye R, Kyratzis IL, Truong YB, Arnold L (2011) Recent advances in nanofibre fabrication techniques. Text Res J 82(2):129–147.
7. Zhang X, Lua Y (2014) Centrifugal spinning: an alternative approach to fabricate nanofibers at high speed and low cost. Polym Rev 54:677–701. https://doi.org/10.1080/ 15583724.2014.935858.
8. Ma PX, Zhang R., Synthetic nano-scale fibrous extracellular matrix, J Biomed Mater Res. 1999 Jul;46(1):60-72.
9. Li W-J, Shanti RM, Tuan RS (2006) Electrospinning technology for nanofibrous scaffolds in tissue engineering. Nanotechnologies Life Sci 9:135–186.
10. Hartgerink JD, Beniash E, Stupp SI.,, Self-assembly and mineralization of peptide-amphiphile nanofibers, Science. 2001 Nov 23;294(5547):1684-8.
11. J. P. Eberhart, Structural and chemical analysis of materials: x-ray, electron and neutron diffraction; X-ray, electron and ion spectrometry; electron microscopy, Wiley Publisher New York, ISBN 0 471 92977 8, 1991.

12. Xiaodong Zou, Sven Hovmöller, and Peter Oleynikov, Electron Crystallography: Electron Microscopy and Electron Diffraction, Oxford University Press, ISBN 9780199580200, 2011.
13. B.D. Cullity, Elements of X-ray Diffraction, Pearson, ISBN 978-0201610918, 2001.
14. Ian Fleming and Dudley Williams, Spectroscopic Methods in Organic Chemistry, Springer Nature Switzerland AG, ISBN 978-3-030-18251-9, 2019.
15. Yadav and Lal Dhar Singh, Organic Spectroscopy, Springer Netherlands, ISBN 978-1-4020-2575-4, 2005.
16. Christopher Jones, Mulloy, Barbara, Thomas, Adrian H., Microscopy, Optical Spectroscopy, and Macroscopic Techniques, Humana Press, ISBN 978-0-89603-232-3, 1994.
17. Gunter Gauglitz and David S. Moore, Handbook of Spectroscopy, Wiley-VCH Verlag GmbH & Co. KGaA ISBN 9783527321506, 2014.
18. Gunter Gauglitz and Tuan Vo-Dinh, Handbook of Spectroscop, Wiley-VCH Verlag GmbH & Co. KGaA, Weinheim, ISBN 3-527-29782- 0, 2003.
19. E W Abel, Organic Spectroscopic Analysis, Royal Society of Chemistry, ISBN 978-0-85404-476-4, 2004.
20. Nakamoto, Infrared and Raman Spectra of Inorganic and Coordination Compounds Infrared and Raman Spectra of Inorganic and Coordination Compounds, Part A: Theory and Applications in Inorganic Chemistry, John Wiley & Sons, Inc., ISBN 9780471743392, 2009.
21. A. D. Cross, and Jones, R. Alan, An Introduction to Practical Infra-red Spectroscopy, Springer US, ISBN 978-1-4899-6274-4, 1969.
22. H. Ishida, Fourier Transform Infrared Characterization of Polymers, Springer US, ISBN 978-1-4684-7778-8, 1987.
23. Philip W. West, Inorganic thermogravimetric analysis, J. Chem. Educ., Vol. 31 (6), 334-339, 1954.
24. P.J. Haines, Thermogravimetry. In: Thermal Methods of Analysis. Springer, Dordrecht, ISBN 978-0-7514-0050-2, 1995.

25. Wesley W. Wendlandt, Inorganic Thermogravimetric Analysis, Inorg. Chem., Vol.3(4), 435-436, 1965.
26. Paul D. Garn, Thermoanalytical Methods of Investigation, ASIN: B0000CMYA0, Academic Press, 1965.
27. Antonín Blazek, Thermal Analysis, London : Van Nostrand Reinhold, ISBN 978-1-2024-0110-3, 1973.
28. J. Goldstein, Newbury, D.E., Joy, D.C., Lyman, C.E., Echlin, P., Lifshin, E., Sawyer, L. and Michael, Scanning electron microscopy and x-ray microanalysis, Springer US, ISBN 978-0-306-47292-3, 2003.
29. J. Peter Goodhew, John Humphreys and Richard Beanland, Electron microscopy and analysis, CRC Press New York, ISBN 9780748409686, 1988.
30. M. H. Loretto, Electron beam analysis of materials, Springer Dordrecht, Dordrecht, ISBN 978-94-010-8944-9, 1994.
31. Pennycook, Stephen J., Nellist and Peter D, Scanning transmission electron microscopy: imaging and analysis. Springer-Verlag New York, ISBN 978-1-4419-7199-9, 2011.
32. David B. Williams, Carter and C. Barry Williams, Transmission Electron Microscopy: A Textbook for Materials Science, ISBN 978-0-387-76500-6, Springer US, 2009.
33. Jeanne Ayache, Luc Beaunier, Jacqueline Boumendil, Gabrielle Ehret and Danièle Laub, Sample Preparation Handbook for Transmission Electron Microscopy, Springer-Verlag New York, ISBN 9781441960870, 2010.
34. J. A Belk, Electron microscopy and microanalysis of crystalline materials. London : Applied Science Publishers, ISBN 978-0853348160, 1979.
35. Marc De Graef, Introduction to conventional transmission electron microscopy, Cambridge University Press, ISBN 9780511615092, 2003.
36. R. F. Egerton, Electron energy-loss spectroscopy in the electron microscope Plenum Press, New York, ISBN 978-1-4419-9582-7, 1996.

37. B. Fultz and J. M. Howe, Transmission electron microscopy and diffractometry of materials, Springer Publisher New York, ISBN 978-3642297601, 2005.
38. Manuel Scimeca, Simone Bischetti, Harpreet Kaur Lamsira, Rita Bonfiglio, and Elena Bonanno1, Energy Dispersive X-ray (EDX) microanalysis: A powerful tool in biomedical research and diagnosis, Eur. J. Histochem., Vol.62(1), 2841-2847, 2018..
39. Steffi Rades, Vasile-Dan Hodoroaba, Tobias Salge, Thomas Wirth, M. Pilar Lobera, Roberto Hanoi Labrador, Kishore Natte, Thomas Behnke, Thomas Grossa and Wolfgang E. S. Unger, High-resolution imaging with SEM/T-SEM, EDX and SAM as a combined methodical approach for morphological and elemental analyses of single engineered nanoparticles, RSC Advances, Vol. 4, 49577-49587, 2014.

Chapter-4

Electrospinning Process Parameters Dependent Investigation of TiO$_2$ Nanofibers

4.1 Introduction

Titanium dioxide (TiO$_2$) is a versatile material which has been well-recognized for its several applications such as cosmetics, protective surface coatings, solar cells, sensors (including chemical, gas& bio), water treatment, paints, batteries and many more. The key demand of this material is its non-toxicity, high chemical stability, bio-degradability and low-cost production. There are three main crystallite phases of TiO$_2$ exists such as anatase and rutile in tetragonal while the brookite in orthorhombic shapes. Out of these phases, anatase and brookite can transform to rutile phase via heat treatment whereas rutile remains stable.

Sol-gel-derived TiO$_2$ possesses the anatase phase however; other phases can be attained by controlling the heat treatment mechanism or by preferring the more acidic solution during the synthesis. The phase transformation could be understood by the two mechanisms; surface energy and precursor chemistry. For the case of anatase phase, the associated surface energy is weaker as compared to others two phases. Further, the geometry of the crystal structure is governed by the precursor chemistry which involves the nucleation and the growth of either phases [1].

One-dimensional (1D) TiO_2 nanostructures such as nanotubes, nanowires, nanobelts, nanofibers etc. have been demanded owing to their fast electron-transport and carrier-collection capability. TiO_2 nanofibers are being investigated as photoanode material in dye-sensitized solar cells (DSSCs) application. However, photoanode based on 1D-TiO_2 nanostructures possess low efficiency mainly due to the weak dye-adsorption associated with the surface morphology. Conversely, by treating the surface with acid, $TiCl_4$ and oxygen plasma one can improve the dye loading. DSSCs based on TiO_2 nanofibers were studied to investigate the influence of the surface treatment [2].

The improved conversion efficiency was noticed in the order 8.59% < 9.33% in accordance with treated photoanodes with acid and oxygen plasma. The morphology of TiO_2 nanofibers has great influence of the calcination temperature. The photocatalytic activity of TiO_2 nanofibers calcined at 500 ⁰C for 3 hours was performed using rhodamine B under visible light irradiation and found satisfactory as compared to nanofibers calcined at temperature 600, 700 ⁰C for 3 hours [3].

Electrospun TiO_2 nanofibers in the range from 194-441 nm have been investigated for the application as the scattering material for the DSSCs [4]. The scattering property was found to be linearly dependent on the diameter and density of the fibers. The photocurrent-voltage characteristics of the DSSCs were evidenced the increased performance which has been attributed to the scattering effect caused by TiO_2 nanofibers. Further, the photocatalytic activity of hydrogen evolution under UV irradiation was studied which endorsed the similar scattering effect as compared to other samples. Various morphologies of electrospun TiO_2 nanofibers have been investigated for DSSCs and photocatalytic applications. By co-axial electrospinning, hollow/tubular TiO_2 nanofibers were prepared and further etching treatment was preferred using sodium hydroxide aqueous solution in order to get the porous morphology of TiO_2 nanofibers [5].

The diameter of the hollow/tubular nanofibers was in the range of 300–500 nm whereas porous nanofibers were in ribbon shape with their width about 200 nm. Brunauer-Emmett-Teller (BET) surface area of the hollow/tubular TiO_2 nanofibers was 27.3 m^2/g, which was almost double of the solid TiO_2 nanofibers; however 106.5 m^2/g surface area was obtained for the porous TiO_2 nanofibers. The morphology study of TiO_2 nanofibers via solution viscosity and electrospinning process parameters has been investigated [6]. The low viscous solution with the high ratio of precursor solution and glacial acetic acid resulted in the beaded morphology of TiO_2 nanofibers.

The optimized process parameters showed the smooth nanofibers with their average diameters 148±79 nm. Photocatalytic activity of mesoporous TiO_2 nanofibers prepared by electrospinning process followed by the solvothermal treatment has also been reported [7]. The additional solvothermal process was found useful to crystallize TiO_2 by arranging closely packed grains which could improve the adsorption of CO_2. As a consequence, enhanced photocatalytic behavior of TiO_2 was noticed which has been regarded as the improved adsorption and the charge separation influenced by the solvothermal process. The conductivity of TiO_2 nanofibers is low owing to its higher resistivity. Therefore, doped-TiO_2 nanofibers have been investigated for the enhancement of conductivity. In addition, enhanced conductivity of TiO_2 nanofibers have been reported by treating with potassium hydroxide which could alter the insulating behavior into conductivity due to their reduced resistivity [8].

This behavioral change of TiO_2 makes this material apposite for supercapacitor application. Potassium hydroxide treated TiO_2 showed the abrupt increase in the magnitude of the capacitance value which was about 1500 times the pristine. Gold doped-TiO_2 nanofibers were prepared and their photocatalytic activity was studied [9]. An enhanced photocatalytic activity was observed with gold-doped TiO_2 nanofibers. This result was attributed to the formation of the Schottky-barrier at the junction of gold-TiO_2 which prevented the carriers recombination and hot electron generation.

In similar way, electrospun silver-TiO_2 nanofibers were reported to study the influence of silver concentration [10]. With the increased doping concentration of silver, the diameter of fibers was found to be increased while photoluminescence intensity was weaker. The antibacterial study was performed with the pathogenic bacteria and found improved as compared to pure-TiO$_2$ nanofibers. The enhanced antibacterial activity has been attributed to the silver doping along with large surface area of the prepared nanofibers. In another approach, graphitic carbon nitride nanosheets hybridized nitrogen-doped titanium dioxide nanofibers in-situ fabrication has been reported by electrospinning process [11]. The prepared hybrid nanofibers were mesoporous structure and its partial decomposition was found responsible for the doping of nitrogen in the bulk-TiO$_2$. As maximum as 8,931.3 µmolh^{-1}g^{-1} photocatalytic H$_2$ production rate was achieved which has been associated with the graphitic-C$_3$N$_4$ nanosheet hybridized N-doped TiO$_2$ nanofibers. This hybrid material was found promising for the improved light absorption and the electron-transport mechanism. Electrospun silver/TiO$_2$ nanofibers were studied for the photocatalytic activity prepared at different sintering temperatures and silver concentrations [12].

The nanofibers calcined at temperature 400 °C was about 120 nm in diameter with 20 nm particles size which showed about 71 % degradation rate of methylene blue. The porous and uniform TiO$_2$/g-C$_3$N$_4$ (graphitic-carbonnitride composite) nanofibers prepared by electrospinning method have been reported [13]. The diameter was found to be 100-150 nm after calcination of nanofibers at temperature 550 °C.

To study the photocatalytic activity of the prepared nanofibers, the degradation of rhodamine B dye under sunlight was evaluated. The photocatalytic activity was found to be increased which has been associated with the hetero-junction $TiO_2/g-C_3N_4$ and was promising for the charge transport mechanism along with the prevention of charge recombination. Over the randomly aligned nanofibers, unidirectional nanofibers possess enhanced optical and mechanical properties with their high degree of crystallinity. These unidirectional grown nanofibers favor the better transport of charge-carriers and therefore, enhanced the performance of the devices [14,15]. A study of electrospun TiO_2 nanofibers by employing the modified aluminum collector of two-pieces has been reported which endorsed the unidirectional growth of nanofibers [16]. Further, this study was explored for the tuning of nanofibers diameter by controlling the tip-collector distance and the applied voltage.

Various techniques such as electrospinning, drawing, template synthesis, phase separation and self-assembly are available for the preparation of polymer-based nanofibers. Among aforementioned techniques, an electrospinning fabrication technique is recognized to be promising for the growth of continuous fibers due to its easy and cost-effective process. Electrospinning system consists of three main parts; 1) metal collector (drum/plate/disk etc.), 2) syringe pump with metal tip/needle and, 3) dc high voltage power supply. The morphology of the electrospun fibers is significantly governed by the process parameters. These process parameters are the applied dc voltage, solution/gel flow rate, distance metal tip-collector and polymer concentration. The diameter of nanofibers decreases with the increase of applied dc voltage and the distance tip-collector. However, these conditions are valid for enough viscous solution otherwise it produces beads/particles by electrospraying rather than electrospinning process. In similar way, the reduced solution flow rate and the polymer concentration yields thinner nanofibers. In addition, ambient environment like humidity and temperature have their significant role for the preparation of continuous nanofibers without any defect.

This work reports the electrospinning fabrication and characterization of TiO_2 nanofibers. Various electrospinning process parameters such as applied voltage, distance tip-collector, solution flow rate and polymer (PVP) concentration are studied. The prepared nanofibers was investigated by X-ray diffraction (XRD), scanning electron microscopy (SEM), energy dispersive X-ray spectroscopy (EDS), thermogravimetric-differential thermal analysis (TG-DTA) and Fourier transform infrared spectroscopy (FTIR). XRD pattern of TiO_2 nanofibers evidenced the presence of mixed phases of anatase and rutile. TG/DTA investigation showed the characteristic peaks correspond to the heating behavior of TiO_2-PVP mat. FTIR investigation endorsed a vibration peak at 660 cm^{-1} associated with the characteristic Ti-O-Ti bond.

With the optimized process parameters, TiO_2 nanofibers diameter was found to be reduced to 74 nm as compared to first sample prepared with diameter 343 nm. Furthermore, these nanofibers were employed as the photoanode material for the preparation of dye-sensitized solar cell (DSSC) and photovoltaic study is evaluated. Section 4.2 describes the experimental details of the electrospinning process. The characterized results have been discussed in section 4.3. Finally, section 4.4 presents the summary of the work.

4.2 Experimental Details
Materials

For the preparation of TiO_2 nanofibers, titanium tetraisopropoxide (TTIP, Sigma-Aldrich), acetic acid solution (Sdfine), polyvinyl pyrolidone (PVP, Mw=1,300,000, Sigma-Aldrich) and methanol (Fisher Scientific) were procured and used without any further purification.

Electrospinning Setup

The electrospinning setup is illustrated in **figure 4.2**.1 It is mainly consists of collector drum, dc power supply and syringe pump. These all mechanisms are assembled in a fume hood whereas its front panel shows the various controls like speed of collector-drum, applied dc voltage, spin rate and flow rate. The right-hand side top image depicts the enlarged image of the collector drum while fiber Taylor cone formation can be observed in the bottom image.

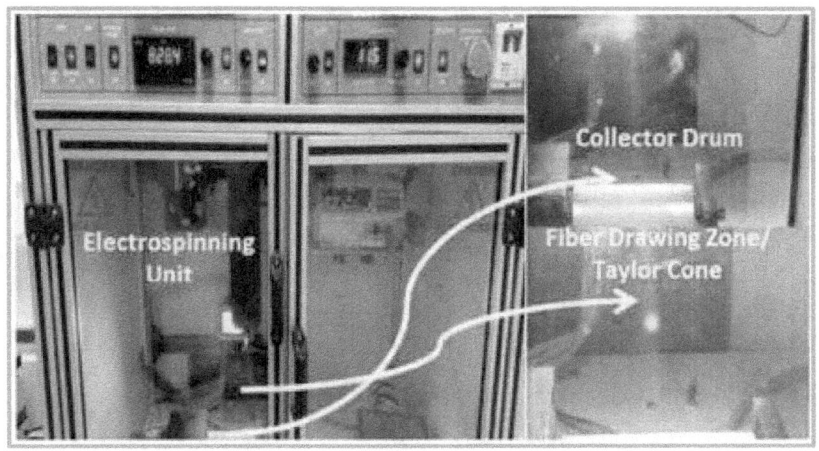

Figure 4.2.1 Electrospinning setup for the preparation of nanofibers.

Methods

Before the electrospinning process, the sol-gel synthesis was performed to get enough viscous solution. At first, 0.6 ml TTIP precursor was vigorously stirred in 10 ml methanol. After 5 min, 4 ml glacial acetic acid was added in TTIP solution and kept for 30 min stirring at room temperature. Later, 1.12 g PVP was dissolved in the above solution and stirred for 3 hr.

The prepared solution was found transparent and enough viscous which was loaded in syringe. For the first electrospinning process, the flow rate and the distance tip-collector were fixed to 3 ml/hr and 10 cm respectively whereas the applied voltage was maintained to 12 kV.

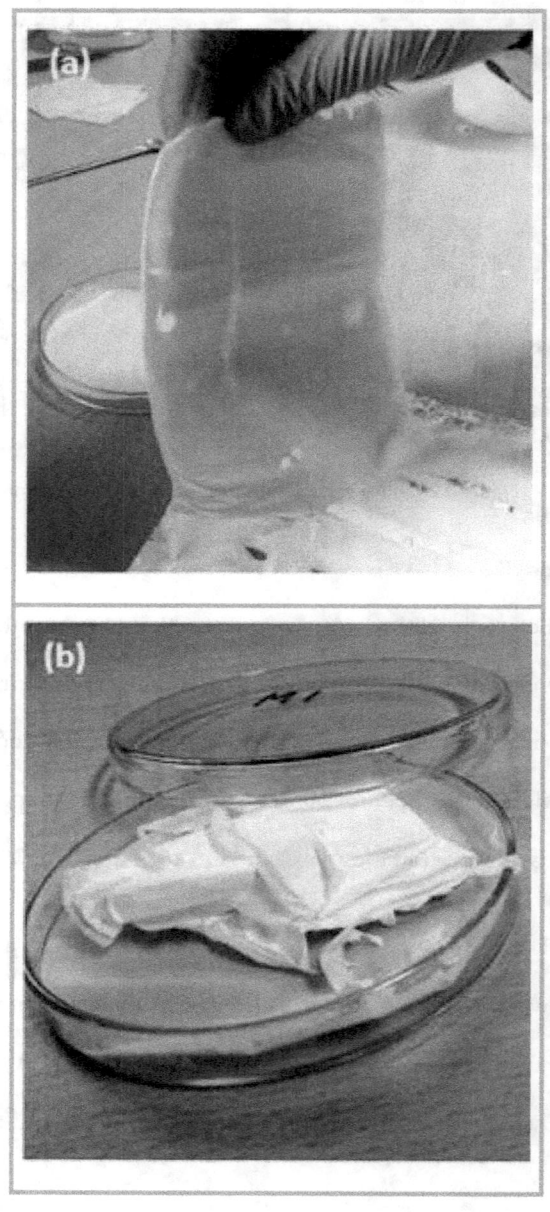

Figure 4.2.2 (a)&(b) Peeling of as-prepared electrospun TiO$_2$-PVP mat on aluminum foil fig.(a) and collected sample fig.(b).

The images of peeling of as-prepared TiO$_2$-PVP mat and the collected one are shown in **figure 4.2.2(a)** and **4.2.2(b)** respectively. Later, optimization of the process parameters was performed by varying the applied voltage, distance tip-collector, flow rate and the polymer concentration.

Characterization

The electrospun TiO$_2$ nanofibers were characterized to examine the phase and crystallinity using X-ray Diffraction (XRD, Bruker AXS D8 Advance, Germany), the qualitative and quantitative analysis using Fourier-transform infrared spectroscopy (FTIR, Shimadzu, Japan), heating behavior of TiO$_2$-PVP mat using thermogravimetric differential thermal analysis (TG-DTA, DTG-60H, Shimadzu), surface morphology study using scanning electron microscope (SEM, JSM-6360, USA) and the compositional chemical elementals investigation using EDS attached to SEM.

4.3 Results and Discussion

4.3.1. XRD Analysis

X-ray diffraction (XRD) measurement was carried out to investigate the crystalline nature of TiO$_2$ nanofibers. **Figure 4.3.1** depicts the XRD patterns of TiO$_2$-PVP mat and TiO$_2$ nanofibers calcined at 450 ^{0}C for 3 h. Before calcination, no appearance of diffraction peak in the XRD pattern indicates the amorphous nature of TiO$_2$-PVP mat. Inversely, various characteristic diffraction peaks can be observed for the case of calcined TiO$_2$ nanofibers at 450 ^{0}C for 3 h.

The XRD pattern indicates the mixed anatase and rutile phases which were assigned to JCPDS#21-1272 and JCPDS#21-1276 respectively. The highest intensity diffraction peak at 2θ= 25 of the plane (101) corresponds the anatase phase of TiO_2. The anatase peaks were originated from the lattice planes at 2θ values, $25^0=d_{101}$, $48^0=d_{200}$, $54^0=d_{211}$, $62^0=d_{204}$ and $74^0=d_{107}$ while rutile peaks at $27^0=d_{110}$, $41^0=d_{200}$, $44^0=d_{210}$ and $69^0=d_{301}$.

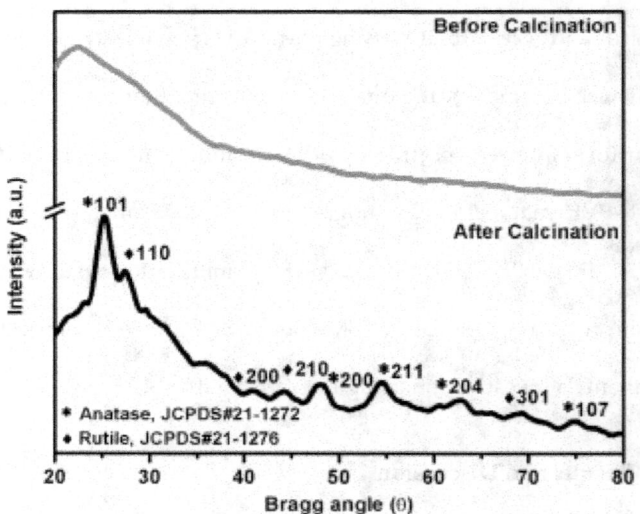

Figure 4.3.1 X-ray diffraction patterns of electrospun TiO_2-PVP mat and calcined TiO_2 nanofibers.

The Scherrer's formula was employed to estimate the crystallite size of the calcined TiO_2 nanofibers. The Scherrer's equation is represented as $d = k\lambda/\beta cos\theta$, where d is the crystalline size, $k=0.89$ is a constant dependent on the crystalline shape, λ is the X-ray wavelength at $1.54056\ A°$ for CuKa, β is the full width at half-maximum intensity, and θ is the Bragg angle. The estimated crystallite size was found to be *8.2 nm* corresponds to the most predominant diffraction peak (101) of anatase phase.

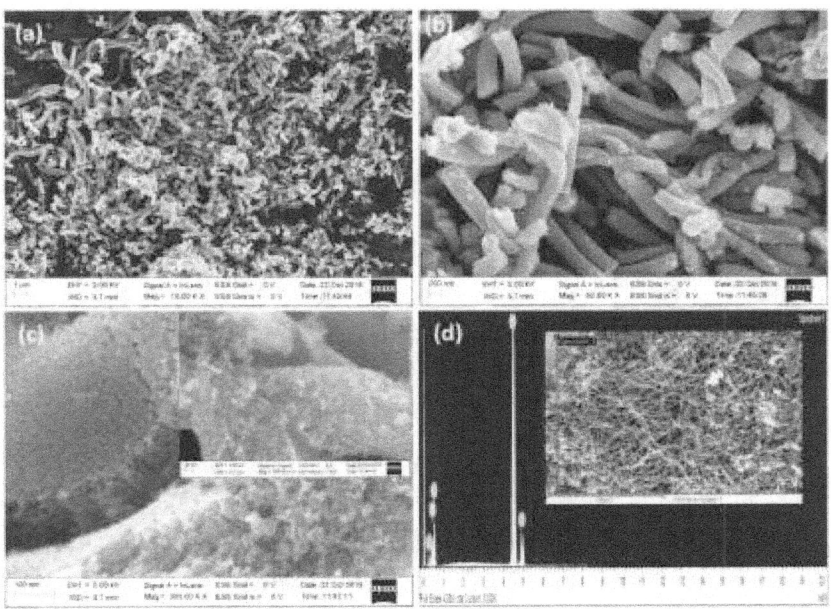

Figure 4.3.2(a),(b),(c),(d),Surface morphology of calcined TiO$_2$ nanofibers at scale 1µm fig.(a), 200 nm fig.(b), 100 nm fig.(c) and EDX spectra fig.(d).

4.3.2. Surface morphology and EDX.

Figure 4.3.2(a),(b),(c),(d) depicts the surface morphology of TiO$_2$ nanofibers calcined at *450 ⁰C* for *3 hr*. We can observe the randomly aligned and smooth morphology of TiO$_2$ nanofibers at scale 1µm as shown in **figure 4.3.2(a)** and at scale 200 nm in **figure 4.3.2(b)**. The diameter of TiO$_2$ nanofibers was found to be in the range of *244-343 nm*. **Figure 4.3.2(c)** endorses the TiO$_2$ particulate at *100 nm* scale while inset image was recorded at *20 nm* scale. TiO$_2$ nanoparticles diameter was found to be *11 nm* as estimated by using jImage open source software. The elemental composition of Ti and O were also confirmed by EDS measurement as shown in **figure 4.3.2(d)**.

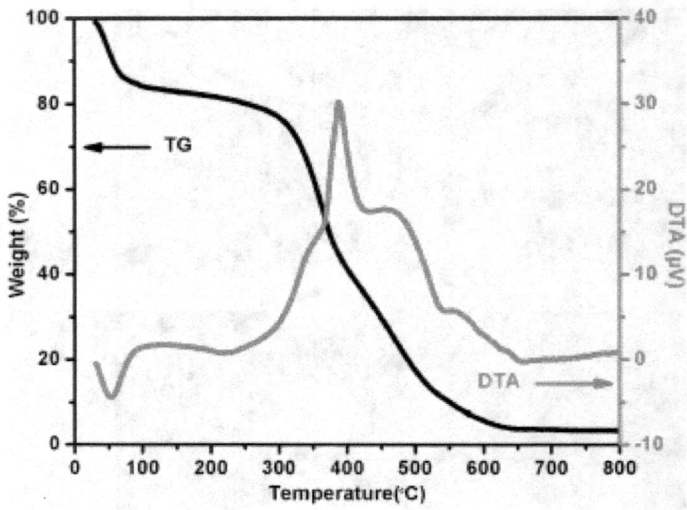

Figure 4.3.3. TG/DTA graph of electrospun TiO$_2$-PVP mat.

4.3.3. TG/DTA Analysis.

To investigate the compositional changes during thermal treatment, thermogravimetric differential thermal analysis (TG/DTA) of TiO$_2$-PVP composite nanofibers was carried out. As depicted in **figure 4.3.3**, the TGA curve endorses the three stages of weight loss. The first region 0-100 °C corresponds to evaporation of residual solvents including desorption of water contents in the sample. The second weight loss is up to 76 % in the temperature region from 100-300 °C indicating the removal of polymer contents from TiO$_2$-PVP mat. However, the third steep weight loss is observed in the region 300-550 °C which reveals the maximum degradation of the polymer with its loss up to 12 %. Referring to DTA curve, an endothermic peak aligned at 80 ^{0}C is assigned to the evaporation of moisture and the solvent while an exothermic peak at 390 ^{0}C can be observed which is attributed to the decompositions of metal hydroxide and polymer. The peak at 500 ^{0}C represents the phase transformation from anatase to rutile [17-20]. However, no mass loss was observed after 640 ^{0}C.

This work is limited to the analyses of optical and structural properties of TiO_2 nanofibers nonetheless the mechanical property is another significant parameter particularly, when the fibers are subjected to mechanical stress during its usage for example as water filter or so. In brief, the as-prepared TiO_2 nanofibers were studied by employing the cantilevered beam bending approach attached to scanning electron microscope [21].

The electrospun nanofibers prepared with the needle and without needle have been studied for Young's modulus and bending strength. By investigations, the electrospun TiO_2 nanofibers prepared with needle showed the uniform morphology with their better mechanical property. In similar way, the mechanical property for the application of nanofiber-reinforced polymer composites has been investigated [22].

For the sample under the test, the hooking and elongation were precisely controlled and the proposed approach was found suitable regardless the both end gripping of nanofibers. The nanofibers based on various diameters have been characterized however, the small diameter based nanofibers showed the better mechanical strength and therefore, it was suggested as the nano-reinforcement for the composite materials.

4.3.4. FESEM Analysis.

Further, we have prepared various samples of TiO_2 nanofibers for the optimization of process parameter such as the applied voltage, the distance tip-collector, the flow rate and the PVP or polymer concentration. Accordingly, **Figures from 4.3.4(a),(b),(c),(d) to 4.3.7(a),(b),(c),(d)** depicts the SEM images of TiO_2 nanofibers and their analyses are presented in further discussion.

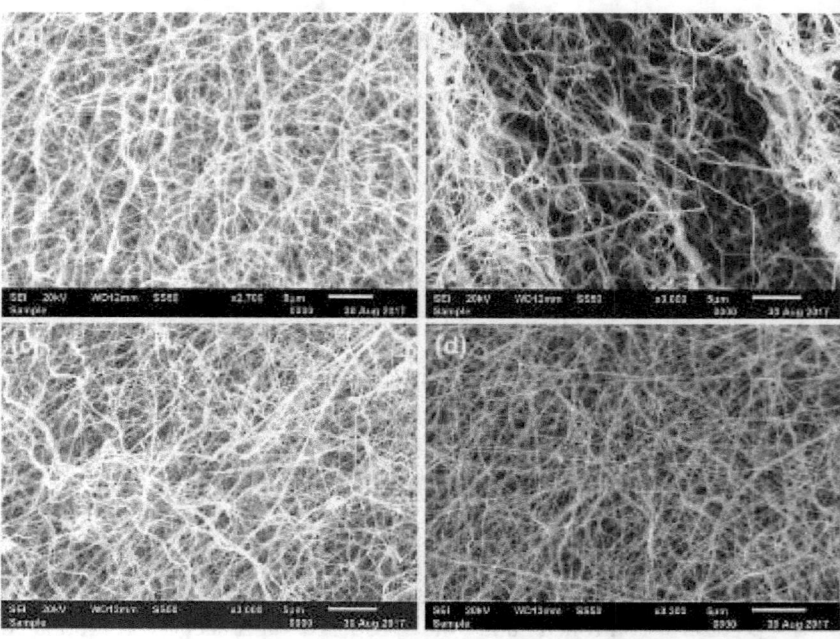

Figure 4.3.4.(a),(b),(c),(d),SEM images of TiO_2 nanofibers prepared at voltages 8, 9, 10 and 11 kV.

Figure 4.3.4(a),(b),(c)and(d) depicts the morphology of electrospun TiO$_2$ nanofibers prepared at voltages 8, 9, 10 and 11 kV respectively. We can observe the smooth and randomly distributed nanofibers with their estimated diameters 293, 226, 189 and 175 nm in accordance with the applied voltages 8, 9, 10 and 11 kV. The distance from tip-collector, flow rate and the PVP concentration were kept at 10 cm, 1 ml/hr and 1g respectively. As the applied voltage was increased the diameter of nanofibers was found to be reduced from 293 nm to 175 nm. Here, we can understand that an optimal applied voltage is important to evaporate the solvent faster while stretching the fiber towards the collector.

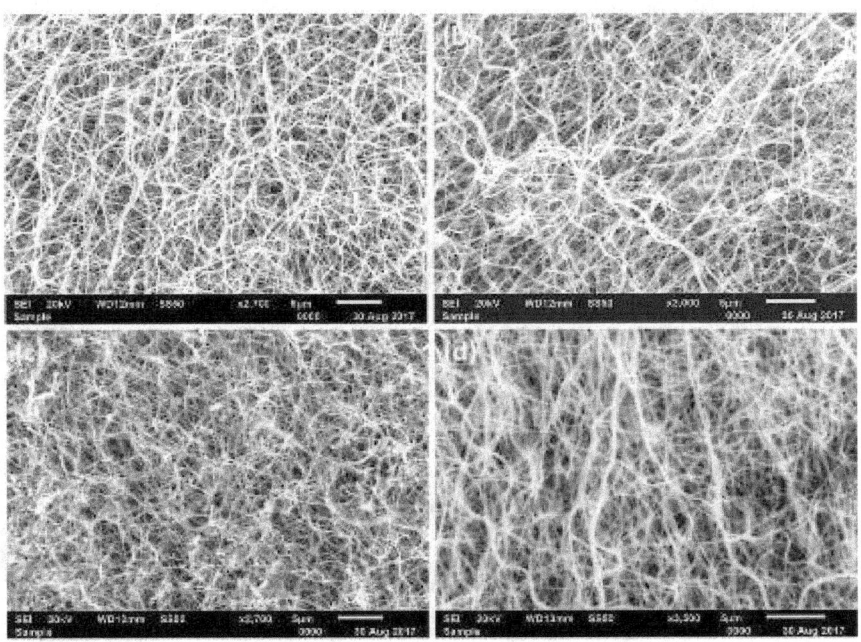

Figure 4.3.5 (a),(b),(c),(d),SEM images of TiO$_2$ nanofibers prepared at distances 8, 10, 12 and 14 cm.

The surface morphology of electrospun TiO$_2$ nanofibers prepared at various tip-collector distances 8, 9, 12 and 14 cm are illustrated in **figure4.3.5 (a)**, **4.3.5 (b)**, **4.3.5 (c)** and **4.3.5(d)** respectively. Here, the parametrical values of the applied voltage, flow rate and the PVP concentration were 10 kV, 1 ml/hr and 1g respectively. The average diameters were found to be 259, 189, 167 and 147 nm corresponding to the tip-collector distance 8, 9, 12 and 14 cm respectively. This analysis reveals that the distance from the tip-collector is a prominent parameter which directly affects the evaporation time required to the solvent and as a result, the fast evaporation process leads to thinner fibers.

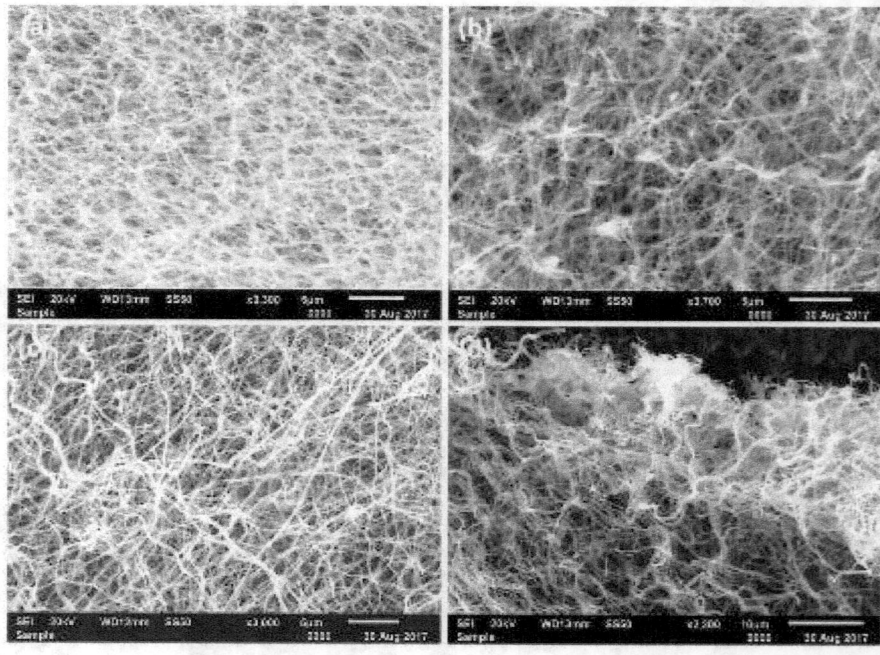

Figure 4.3.6(a),(b),(c),(d),SEM images of TiO$_2$ nanofibers prepared at flow rates 0.6, 0.8, 1.0 and 1.2 ml/hr.

We can observe the formation of randomly distributed smooth nanofibers in **figure** 4.3.6(a),(b),(c),(d), as a function of solution flow rate. The average diameters were found to be 111, 155, 189 and 247 nm in accordance with the flow rates 0.6, 0.8, 1.0 and 1.2 ml/hr. During this optimization process, the applied voltage, tip-collector distance and the PVP concentration were maintained to 10 kV, 10 cm and 1g respectively. Here, the reduced diameter (111 nm) of nanofibers shows the importance of an optimum flow rate of solution to get the thinner fibers without any beads.

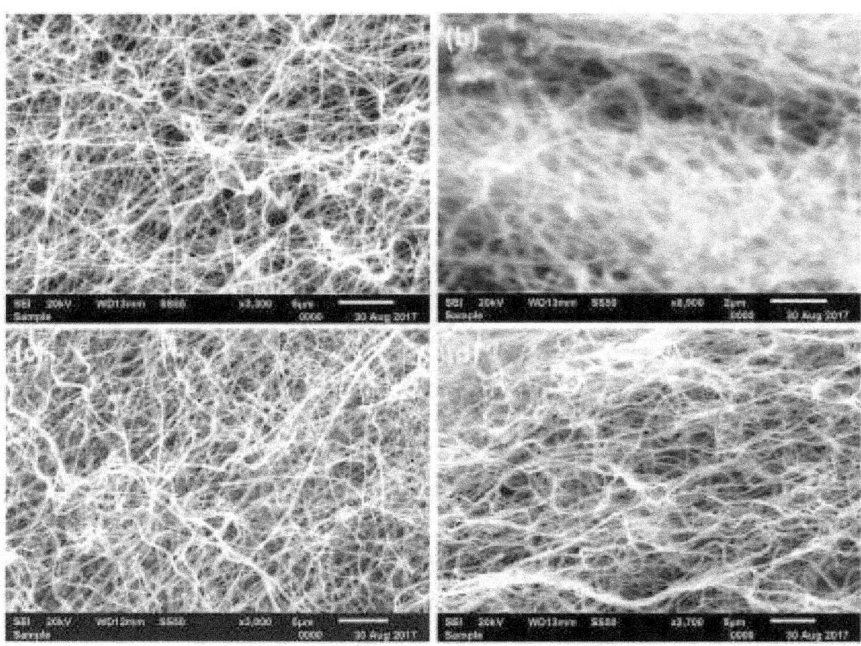

Figure 4.3.7 (a),(b),(c),(d), SEM images of TiO$_2$ nanofibers prepared at PVP concentrations 0.6, 0.8, 1.0 and 1.2 g.

Finally, PVP concentration was optimized while keeping the applied voltage, tip-collector distance and flow rate to 10 kV, 10 cm and 1 ml/hr respectively. The SEM images of PVP concentration is shown in **figure 4.3.7 (a),(b),(c),(d)**. The diameter of the nanofibers were found to be 102, 152, 189 and 284 nm with respect to the PVP concentration 0.6, 0.8, 1.0 and 1.2 g. Depending upon the optimal PVP concentration, the thinner nanofibers can be obtained as it is observed in **figure 4.3.7(a)**. The influence of the four process parameters such as the applied voltage, distance tip-collector, flow rate and the PVP concentration is summarized in **Figure 4.3.8.(a),(b),(c),(d)**

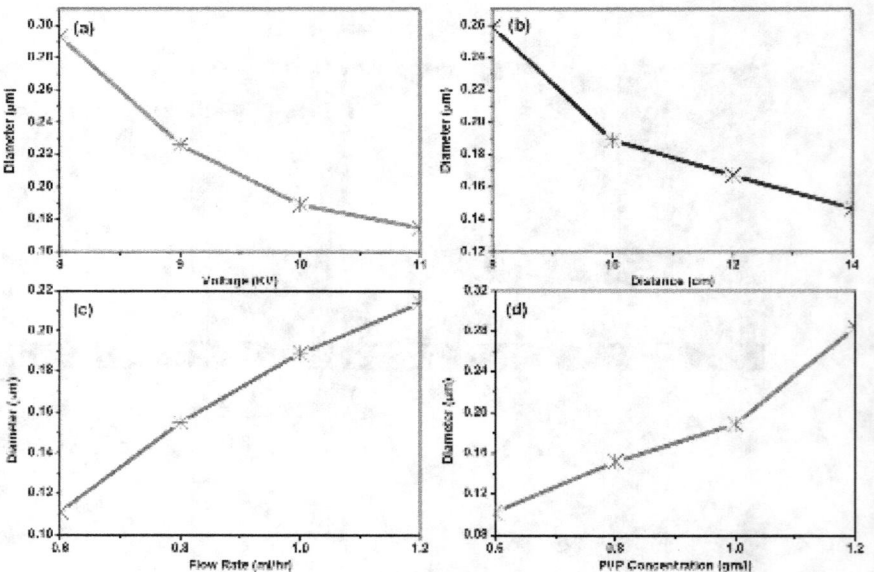

Figure 4.3.8.(a),(b),(c),(d),Diameter of TiO$_2$ nanofibers as a function of applied voltage, distance tip-collector, flow rate and the PVP concentration.

In general, the applied voltage must satisfy the minimum required voltage that is exceeding the threshold voltage for the ejection charged jets from the Taylor cone so that electrospinning process gets started. An increased voltage enhances the electrostatic force on the solution which causes the stretching of jet and hence, leads to thin nanofibers. Accordingly, a decrease in nanofibers diameter from 293-175 nm can be noticed in **figure 4.3.8(a)**. The morphology of the nanofibers has great influence of the applied voltage as observed in SEM images shown earlier in **figure 4.3.4** (a),(b),(c),(d).

In brief, an optimum voltage is recommended to have thin fibers whereas higher voltage results in beads formation. The distance from tip-collector is another parameter which influences the surface morphology and diameter of the nanofibers as observed in **figure 4.3.5** (a),(b),(c),(d). Therefore, a minimum distance is required for the evaporation of the solvent that too before the fiber reaches to the collector during the electrospinning process. As the distance tip-collector increases, accordingly the fibers diameter reduces from 259-147 nm as shown in **figure 4.3.8(b)**. A large distance produces the thin fibers however; beads formation can be prevented by avoiding too near or too far distance from the tip-collector.

The flow rate is a significant parameter which deals with the polymer solution transfer rate and its speed. Generally, sufficient time is needed for the solvent evaporation which can be attained by choosing a small flow rate. An optimal flow rate is fine for the smooth fiber preparation however, a high flow rate yields beaded fibers as it gets lesser time for the solvent evaporation. **Figure 4.3.8(c)** depicts the increment in fibers diameter from 111-214 nm as a function of flow rate. In general, the high flow rate decreases the charge density which produces the larger diameter. It means an increase in feed rate yields corresponding increase in the diameter of the fibers. At last, PVP concentration is another important parameter which yields thinner fibers from 284-102 nm by decreasing the concentration of the polymer as evidenced in **figure 4.3.8(d)**.

An optimum concentration of polymer yields smoother and thinner fibers without any beads as observed in **figure4.3.7(a),(b),(c),(d)**. In this way, the diameter of the electrospun nanofibers can be tuned by optimizing the applied voltage, the distance tip-collector, the flow rate and the polymer concentration. An electrospinning approach is recognized as an easy and low-cost process whereas the reproducibility of the nanofibers can be attained by opting optimal values of these parameters.

After optimizing the process parameters, we have obtained the optimal values of applied voltage 11 kV, distance tip-collector 14 cm, flow rate 0.6 ml/hr and PVP concentration 0.6 g. Finally, we have performed an electrospinning process while maintaining the above parametrical values.

Figure 4.3.9(a)&(b), shows the SEM image and EDS spectra of TiO_2 nanofibers prepared by keeping the optimal values of the process parameters. We can observe the continuous and randomly oriented nanofibers with their average diameter 74 nm in **figure 4.3.9(a)**. This reduced diameter of the nanofibers is attributed to the optimized parameters. The elemental composition of Ti and O is also endorsed in the EDS spectra depicted in **figure 4.3.9(b)**.

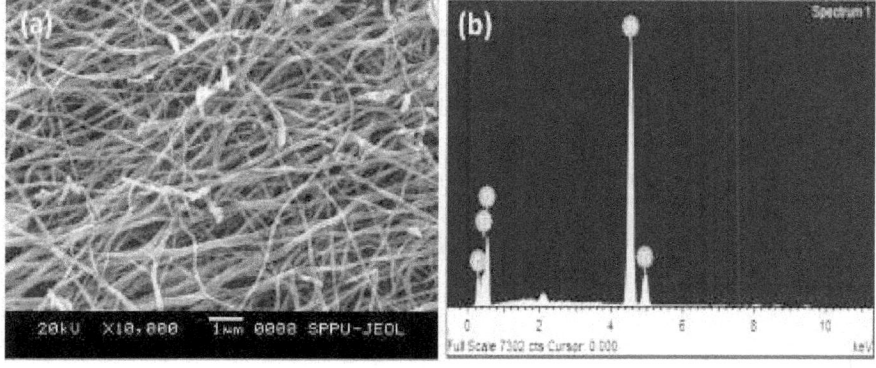

Figure 4.3.9(a)&(b) SEM image fig.(a) and EDS spectra fig.(b) of TiO_2 nanofibers prepared at optimized electrospinning process parameters respectively.

Further, the analysis of chemical bonding and compositions were performed by FTIR measurement in the range of 4000-400 cm^{-1} as shown in **figure 4.3.10(a)**. The various vibration peaks were observed and found good in agreement with reported literatures [17-19]. Accordingly, the peak at 3432 cm^{-1} represents the O-H stretching vibration associated with the absorbed water. An asymmetric peak at 2930 cm^{-1} corresponds to the stretching vibration of C-H group while another peak at 2850 cm^{-1} is attributed to the symmetric stretching mode of C-H.

The stretching mode of C=O vibration is observed at 1640 cm^{-1} whereas peak at 1457 cm^{-1} endorses the bending vibration of CH_2 group. The C-N group of polymer associated to the asymmetric stretching vibration is assigned at 1113 cm^{-1} and the vibration peak at 660 cm^{-1} is attributed to the characteristic Ti-O-Ti bonds.

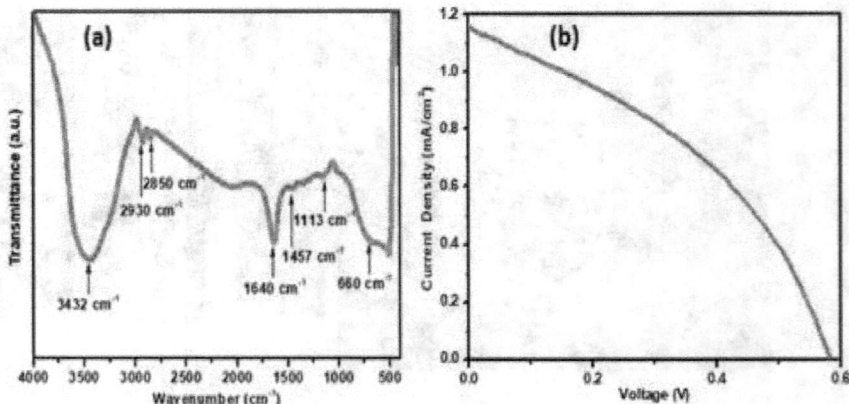

Figure 4.3.10(a)&(b), FTIR spectra of electrospun TiO$_2$ nanofibers prepared fig.(a) and current density-voltage characteristics of DSSC based on TiO$_2$ nanofibers fig.(b).

This work is limited to the investigation of electrospun TiO$_2$ nanofibers via various process parameters however, we have extended the testing of TiO$_2$ nanofibers as the photoanode material of dye-sensitized solar cell (DSSC). In brief, 0.15 g of TiO$_2$ nanofibers of diameter 74 nm was mixed in 0.25 ml ethanol, 0.25 ml acetic acid and 0.25 ml of PVP to get slurry.

The fluorine-doped (FTO) glasses of 1 cm^2 were used as the electrodes after ultrasonically cleaned in ethanol and acetone simultaneously. By doctor blade method, TiO$_2$ paste was rolled-on FTO glass within the active area 0.25 cm^2. After coating, it was dried at temperature 60 ºC and finally sintered at 500 ºC for 30 min. Later, the photoanode was soaked in rhodamine B dye for 12 h and then used after washing with ethanol and de-ionized water.

For the preparation of counter electrode, the platinum paste (Plastisol T, Solaronix) was rolled on the FTO glass using doctor blade method. The assembly of DSSC was done by clipping the counter electrode on top of the dye loaded photoanode with offsetting the electrodes in the opposite direction for the alligator wires connections. To prevent the shorting between the counter-electrode and the photoanode, an insulating film was placed which was kept open at one offset edge for the insertion of electrolyte solution.

After dropping electrolyte solution, the binder clips were slightly opened and closed in order to allow the solution in the active area of the cell. For the measurement of current density-voltage characteristic of DSSC, Keithley 2420 power source and white LED source with 80 mW/cm^2 illumination was employed.

Figure 4.3.10(b) shows the current density-voltage characteristic of DSSC which shows its conversion efficiency 0.38 % with open circuit voltage 0.58 V, short-circuit current 1.15 mA/cm^2 and 0.39 fill factor. DSSC performance can be enhanced by using thinner nanofibers as the host material with optimizing the preparation of photoanode [23]. Besides, photoanode based on the hybrid structure of TiO_2 nanoparticles and nanofibers as the scattering have been reported which showed the enhanced photovoltaic performance [24,25].

4.4. Conclusions

The electrospinning fabrication and characterization of TiO_2 nanofibers have been reported. XRD measurement endorsed the mixed anatase and rutile phases. TG/DTA investigation evidenced the characteristic peaks of TiO2/PVP mat. By FTIR study, the vibration peak at 660 cm^{-1} corresponding to characteristic Ti-O-Ti bond is observed.

SEM measurement showed the randomly distributed nanofibers with their diameter in the range 244-343 nm before the optimization. Furthermore, optimization of various electrospinning process parameters was performed. An increased applied voltage and the distance tip-collector led to preparation of thinner nanofibers from 293-175 nm and 259-147 nm respectively. While the decreased flow rate and PVP concentration evidenced the further thin nanofibers from 214-111 nm and 284-102 nm respectively.

Furthermore, the optimized values of electrospinning process parameters could reduce the diameter of TiO_2 nanofibers to 74 nm. Using thinner TiO_2 nanofibers, DSSC photoanode was fabricated and photovoltaic performance was evaluated. Finally, this study is helpful to optimize the electrospinning process parameters for the preparation of thin/thick nanofibers and their reproducibility using an easy and in-expensive method.

References

1. D Reyes-Coronado, G Rodrıguez-Gattorno, M E Espinosa-Pesqueira, C Cab, RdeCoss and G Oskam, Phase-pure TiO_2 nanoparticles: anatase, brookite and rutile, Nanotechnology Vol. 19 145605, 10pp (2008).

2. Y. Li, J.N. Ding, N.Y. Yuan, L. Bai, H.W. Hu, X.Q. Wang, The influence of surface treatment on dye-sensitized solar cells based on TiO_2 nanofibers, Materials Letters, Vol. 97, 74-77 (2013).

3. Jing Li, Hui Qiao, Yuanzhi Du, Chen Chen, Xiaolin Li, Jing Cui, Dnt Kumar, and Qufu Wei, Electrospinning Synthesis and Photocatalytic Activity of Mesoporous TiO_2 Nanofibers, The Scientific World Journal. Article ID 154939, 7 pages, doi:10.1100/2012/154939 (2012).

4. Yuan-Lian Chen, Yi-Hao Chang, Jow-Lay Huang, Ingann Chen and Changshu Kuo, Light Scattering and Enhanced Photoactivities of Electrospun Titania Nanofibers, J. Phys. Chem. C 2012, 116, 3857-3865 (2012).

5. Guangfei He, Yibing Cai, Yong Zhao, Xiaoxu Wang, Chuilin Lai, Min Xi, Hao Fong and Zhengtao Zhu, Electrospun anatase-phase TiO_2 nanofibers with different morphological structures and specific surface areas, Journal of Colloid and Interface Science, Vol. 398, 103–111 (2013).

6. Soraya Mirmohammad Sadeghi, Mohammadreza Vaezi, Asghar Kazemzadeh and Roghayeh Jamjah, Morphology enhancement of TiO_2/PVP composite nanofibers based on solution viscosity and processing parameters of electrospinning method, 2018 J. Appl. Polym. Sci., Vol. 135, 46337, 11pg, (2018).

7. Junwei Fu, Shaowen Cao, Jiaguo Yu, Jingxiang Low and Yongpeng Lei, Enhanced photocatalytic CO-reduction activity of electrospun mesoporous TiO_2 nanofibers by solvothermal treatment, Dalton Trans, Vol. 43, 9158-9165 (2014).

8. X. He, C.P.Yang, G.L.Zhang, D.W.Shi, Q.A.Huang, H.B.Xiao, Y.Liu and R.Xiong, Supercapacitor of TiO_2 nanofibers by electrospinning and KOH treatment, Materials and Design, 106, 74–80 (2016).

9. Tao Wang, Yu Zhang, Yong Wang, Jinxin Wei, Ming Zhou, Zhengmei Zhang and Qi Chen, One-Step Electrospinning Method to Prepare Gold Decorated on TiO_2 Nanofibers with Enhanced Photocatalytic Activity, Journal of Nanoscience and Nanotechnology, 18, 3176–3184 (2018).

10. M. A. Kudhier, R. S. Sabry, Y. K. Al-Haidarie and M. F. AL-Marjani, Significantly enhanced antibacterial activity of Ag-doped TiO_2 nanofibers synthesized by electrospinning, Materials Technology, DOI: 10.1080/10667857.2017.13967 (2017).

11. Cheng Han, Yingde Wang, Yongpeng Lei, Bing Wang, Nan Wu, Qi Shi and Qiong Li, In situ synthesis of graphitic-C3N4 nanosheet hybridized N-doped TiO_2 nanofibers for efficient photocatalytic H2 production and degradation, Nano Research, Vol. 8, 1199-1209 (2015).

12. Hong Ying Dong, Qing Hong Sun, Ting Ting Zhang, Qi Ren and Wen Ma, Synthesis and Photocatalytic Activity of Ag Doped TiO_2 Nanofibers, Materials Science Forum, Vol. 913, 1027-1032 (2018).

13. Qian Tang, Xianfeng Menga, Zhiying Wang, Jianwei Zhou, HuaTang, One-step electrospinning synthesis of TiO_2/g-C3N4nanofibers with enhanced photocatalytic properties, Applied Surface Science, Vol. 430, 253-262 (2018).

14. Jalili, R., Morshed, M., Ravandi, S.A.H. Fundamental Parameters Affecting Electrospinning of PAN Nanofibers as Uniaxially Aligned Fibers. J. Appl. Polym. Sci., 101, 4350–4357 (2006).

15. Kovtyukhova, N.I., Mallouk, T.E., Nanowires as Building Blocks for Self-Assembling Logic and Memory Circuits. Chem. Eur. J. 8, 4354–4363 (2002).

16. Jae-Hun Kim, Jae-Hyoung Lee, Jin-Young Kim and Sang Sub Kim, Synthesis of Aligned TiO_2 Nanofibers Using Electrospinning, Appl. Sci. 8, 309, 10pp (2018).

17. Meiqi Chang, Ye Sheng, Yanhua Song, Keyan Zheng, Xiuqing Zhou and Haifeng Zou, Luminescence properties and Judd-Ofelt analysis of TiO_2: Eu3+ nanofibers via polymer-based electrospinning method, RSC Adv., Vol.6, 52113-52121 (2016).

18. Ho-Hwan Chun and Wan-Kuen Jo, Polymer material-supported titania nanofibers with different Polyvinylpyrrolidone to TiO_2 ratios for degradation of vaporous Trichloroethylene, Journal of Industrial and Engineering Chemistry, Vol. 20, 1010–1015 (2014).

19. Wenkai Chang, Fujian Xu, Xueyan Mu, Lili Ji, Guiping Ma, Jun Nie, Fabrication of nanostructured hollow TiO_2 nanofibers with enhanced photocatalytic activity by coaxial electrospinning, Materials Research Bulletin, Vol.48 2661–2668 (2013).

20. Wiwat Nuansing, Siayasunee Ninmuang, Wirat Jarernboon, Santi Maensiri, Supapan Seraphin, Structural characterization and morphology of electrospun TiO_2 nanofibers, Materials Science and Engineering B, Vol. 131, 147-155 (2006).

21. Mikk Vahtrus, Andris Sutka, Sergei Vlassov, Anna Sutka, Boris Polyakov, Leonid Dorogin, Rünno Lõhmus and Rando Saar, Mechanical characterization of nanofibers using a nanomanipulator and atomic force microscope cantilever in a scanning electron microscope, Materials Characterization, Vol. 100, 98–103 (2015).

22. Kenny Yoonki Hwang, Sung-Dae Kim, Young-Woon Kim and Woong-Ryeol Yu, Mechanical characterization of nanofibers using a nanomanipulator and atomic force microscope cantilever in a scanning electron microscope, Polymer Testing, Vol. 29, 375-380 (2010).

23. I. Jinchu, C.O Sreekala, U.S. Sajeev, K. Achuthan, K.S. Sreelatha, Photoanode Engineering Using TiO$_2$ Nanofibers for Enhancing the Photovoltaic Parameters of Natural Dye Sensitised Solar Cells, Journal of Nano- and Electronic Physics, Vol. 7, 04002 (4pp) (2015).

24. Ji-Hye Lee, Kyun Ahn, Soo Hyung Kim, Jong Man Kim, Se-Young Jeong, Jong-Sung Jin, Euh Duck Jeong, Chae-Ryong Cho, Thickness effect of the TiO$_2$ nanofiber scattering layer on the performance of the TiO$_2$ nanoparticle/TiO$_2$ nanofiber-structured dye-sensitized solar cells, Current Applied Physics, 14, 856-861 (2014).

25. G. S. Anjusree, T. G. Deepak, K. R. Narendra Pai, John Joseph, T. A. Arun, Shantikumar V. Nair and A. Sreekumaran Nair, TiO$_2$ nanoparticles @ TiO$_2$ nanofibers – an innovative one-dimensional material for dye-sensitized solar cells, RSC Adv. 4, 22941-22945 (2014).

Chapter-5

Preparation and Investigation of TiO$_2$/ZnO Composite Nanofibers for Photocatalytic Applications

5.1 Introduction

Titanium oxide (TiO$_2$) and zinc oxide (ZnO) are the equally demanded materials in several applications such as batteries, sensors, photocatalytic /water splitting, dye-sensitized solar cells etc. [1-2]. TiO$_2$ is abundant in nature, non-toxic and easy to handle. The anatase and rutile polymorphs of TiO$_2$ are the well-recognized ones due to their better stability as compared to its other forms. The anatase and rutile TiO$_2$ phases have 3.2 and 3.0 eV optical band gaps respectively. On the other hand, the ZnO material possesses better electrical properties and investigated as the alternate choice in place of the TiO$_2$ due to its almost similar band gap (3.37 eV) with its wurtzite hexagonal phase structure. Therefore, both TiO$_2$ and ZnO have been employed as the photocatalysts [3,4].

Further, for the improved photodegradation, the TiO$_2$/ZnO composite semiconductor materials are found to be promising which boosts the process of electron-hole pair separation under light irradiation. Therefore, TiO$_2$/ZnO hybrid nanomaterials in the form of particles, films, fibers etc. have been well-recognized for the photodegradation study [5,6].

For the preparation of nanofibers based on mono, composite or core-shell materials, an electrospinning technique is a simple, inexpensive and well-recognized one. Chun et al. reported the photodegradation study of dye using electrospun TiO_2/ZnO nanofibers as the catalyst by tuning the anatase-rutile ratio via calcination process at various temperatures. The nanofibers calcined at 650 ºC having ratio anatase (48 %) to rutile (52 %) demonstrated the better photocatalytic efficiency with the rhodamine B dye [7]. Yar et al. presented the photodegradation study using catalysts based on electrospun polyacrylonitrile nanofibers decorated with the TiO_2, ZnO and TiO_2/ZnO composite nanoparticles. The degradation of malachite green dye was performed however; the hybrid nanofibers of $TiO_2/ZnO/PAN$ evidenced the photocatalytic activity two times greater the bare PAN nanofibers [8].

Baek et al. fabricated and investigated the various properties of the electrospun ZnO nanofibers. The choice of calcination temperature showed a significant impact on the ZnO structure and endorsed the improved morphology at the higher temperature. In addition, the choice of higher temperature calcination led to the increased diameter of the ZnO nanofibers [9].

Li et al. reported the improved photocatalytic activity of the electrospun TiO_2/ZnO composite nanofibers. Further, the recycled experiment of the photodegradation showed better photodegradation efficiency and the stability of the catalyst [10]. Araujo et al. demonstrated the photodegradation study of the rhodamine B dye using catalyst based on TiO_2/ZnO hierarchical heteronanostructures of nanorods prepared on electrospun nanofibers. The morphological study revealed the three-dimensional arrangement of ZnO nanorods of hexagonal wurtzite on the TiO_2 nanoporous structure. These prepared nanostructures demonstrated as the efficient photocatalyst which showed about 90 % photodegradation of the dye within the 70 min [11]. Lotus et al. reported the fabrication of the TiO_2/ZnO hybrid nanofibers by electrospinning technique with their diameter in the range from 50-150 nm after the calcination. UV-vis measurement study showed two energy band gaps of TiO_2 at 3.0 and 3.5 eV while X-ray diffraction study exhibited the anatase and rutile phases of the TiO_2 and the wurtzite phase of the ZnO.

This study concluded the electrospinning process as the easy fabrication technique for the TiO_2/ZnO hybrid nanofibers with the controllability of various properties such as structural, optical, morphological, thermal and chemical compositional [12]. Pei et al. explored the preparation of the TiO_2/ZnO nanofibers to study the photocatalytic activity. By varying the zinc acetate concentration, various composite fibers were fabricated and characterized. An optimal quantity of the catalyst was determined to increase the photocatalytic response of the composite fibers; as a result, enhanced dye degradation was noticed under visible light irradiation [13].

Liu et al. prepared the core/shell (ZnO/TiO_2) nanofibers for the photocatalytic application. X-ray diffraction study exhibited the anatase and rutile phases of the TiO_2 while hexagonal wurtzite phase was noticed corresponding for the ZnO. Comparatively, the core/shell (ZnO/TiO_2)

nanofibers showed the red-shift regarded the less activation energy requirement in contrast to individual ZnO and TiO$_2$ nanofibers. As a result, the ZnO/TiO$_2$ core-shell nanofibers evidenced the enhanced photocatalytic activity along with their recyclability [14].

To study the photocatalytic activity, Li et al. presented a distinct experimental process to prepare the heterojunction ZnO/TiO$_2$ composite fibers. The approach was zinc plating on the electrospun TiO$_2$ nanofibers and then the heat treatment. As a result of these processes, the photocatalytic activity of the ZnO/TiO$_2$ nanofibers was found to be reasonably higher than the pure-TiO$_2$ nanofibers [15]. In another work, Hwang et al. employed the electrospun ZnO/TiO$_2$ hybrid nanofibers and studied the antimicrobial activity. A promising antimicrobial activity was investigated in the presence of gram-negative Escherichia coli and gram-positive Staphylococcus aureus under ultra-violet (UV) irradiation and in the dark as well [16].

Chen et al. studied the various properties of the ZnO/TiO$_2$ heterogeneous nanofibers fabricated by electrospinning process and later photocatalytic behavior was demonstrated. These nanofibers exhibited the significant degradation of the rhodamine B dye which was attributed to the reduced photo-induced charge carriers rate, improved usage of UV light and the large contact area of the catalyst [17]. Kanjwal et al. reported the photocatalytic activity of the hydrothermally treated electrospun ZnO/TiO$_2$ nanofibers. The photocatalytic activity of the ZnO nanoparticles prepared by hydrothermal process, electrospun TiO$_2$, ZnO/TiO$_2$ composite nanofibers and the hydrothermally treated electrospun ZnO/TiO_2 composite nanofibers were investigated and compared. The hydrothermally treated ZnO/TiO_2 nanofibers were found to be more efficient which could degrade the methyl red and rhodamine B dyes within 90 and 105 min respectively as compared to 3 h degradation time consumed by the other three catalysts [18].

Liu et al. studied the TiO_2/ZnO composite nanofibers calcined at various temperatures. To improve the photocatalytic property, the ZnO was blended in TiO_2/ZnO composite nanofibers. As a result, the photodegradation of the rhodamine B and phenol were noticed to be 100 and 85 % respectively with the 15.76 wt % ZnO content as the optimal quantity for the enhanced photocatalytic response [19]. Park et al. explored the structural and electrical properties of the electrospun ZnO nanofibers. They reported the annealing temperature as the significant factor for the improved crystallinity of the ZnO nanofibers. The electrical conductivity of the ZnO nanofibers was observed to be inversely proportional to the calcination temperature. In addition, the ZnO nanofibers was tested for the CO gas sensing ability and found to be reliable [20].

This works presents the electrospinning fabrication and characterization of TiO_2/ZnO composite nanofibers by varying the proportion of the TiO_2/ZnO solutions. For the fabrication of TZ composite nanofibers, different proportions of TiO_2/ZnO solutions (1:1, 1:2, 2:1 and 1:3) were preferred. X-ray diffraction (XRD) patterns of the T-sample exhibited the mixed anatase and rutile phases while the wurtzite phase is examined for the Z-sample. The field-emission scanning electron microscopy (FESEM) study evidenced the preparation of continuous and randomly oriented nanofibers.

Transmission electron microscopy (TEM) investigation endorsed the cylindrical morphology of the TZ13-sample with c.a. diameter 230 nm. The lattice d-spacing is estimated to be 0.298 nm which corresponds to the plane (100) of the hexagonal wurtzite ZnO while selected area electron diffraction (SAED) analysis endorsed the polycrystalline nature of the TZ13 composite nanofibers. Finally, photodegradation of the Eriochrome black T dye was performed and the catalyst TZ13 showed the significant photodegradation of the dye as compared to TZ11, TZ12 and TZ21. This boosted photocatalytic activity is ascribed to the synergetic effect of the TiO_2/ZnO composite nanofibers with their distinct morphology. Section 5.2 describes the chemicals and experimental processes for the preparation of nanofibers. The characterized results and photocatalytic investigation have been presented in section 5.3. Finally, section 5.4 summarizes the work.

5.2. Experimental Details

Titanium Tetraisopropoxide (TTIP, Sigma-Aldrich) and Zinc Acetate Dehydrate (Zn(CH3COO)2·2HO, Merck) were used as the Ti and Zn precursors respectively. Catalyst: Acetic Acid (Sdfine), polymer: Polyvinyl Pyrolidone (PVP, Mw=1,300,000, Sigma-Aldrich) and the solvent: Methanol (Fisher Scientific) were used. For the photodegradation study, the Eriochrome Black T (EBT) dye was used.

For the preparation of TiO_2 solution, 0.6 ml TTIP, 10 ml methanol and 4 ml glacial acetic acid were mixed and stirred for 30 min. Later, 0.8 g PVP was added in the above solution under vigorous stirring which was continued till the preparation of the transparent and enough viscous solution. To obtain the ZnO solution, 1.2 g zinc acetate dehydrate was dissolved in 10 ml methanol followed by adding 4 ml acetic acid under stirring. Later, 0.8 g PVP was added in the ZnO solution and stirred for 1 hr to get enough viscous solution. After preparing both the solutions, composite solutions of TiO_2 and ZnO were prepared in the proportional ratio 1:1, 1:2, 2:1 and 1:3. For all the electrospinning experiments, the parameters like dc voltage, solution flow rate and the tip-collector distance were 14 kV, 1 ml/hr and 10 cm respectively.

After preparation, each sample was calcined at 600 ºC for 1 hr and named as TZ11, TZ12, TZ21 and TZ13 corresponding to the 1:1, 1:2, 2:1 and 1:3 proportional ratio of the TiO_2:ZnO solutions. The calcined nanofibers were characterized to investigate the phase and crystallinity using X-ray Diffraction (XRD, Bruker AXS D8 Advance, Germany), the Uv-vis absorbance using UV-Visible Spectrophotometer (UV 1800, Shimadzu, Japan), the surface morphology using field-emission scanning electron microscopy (ZIESS, Germany) and Transmission Electron Microscope (Tecnai G2 20 Twin FEI, Netherlands).

5.3. Results and Discussion

5.3.1. XRD Analysis

X-ray diffraction (XRD) study was performed to examine the phase and crystallinity of the TiO_2 (T) and ZnO (Z) nanofibers as depicted in **figure 5.3.1(a)**.

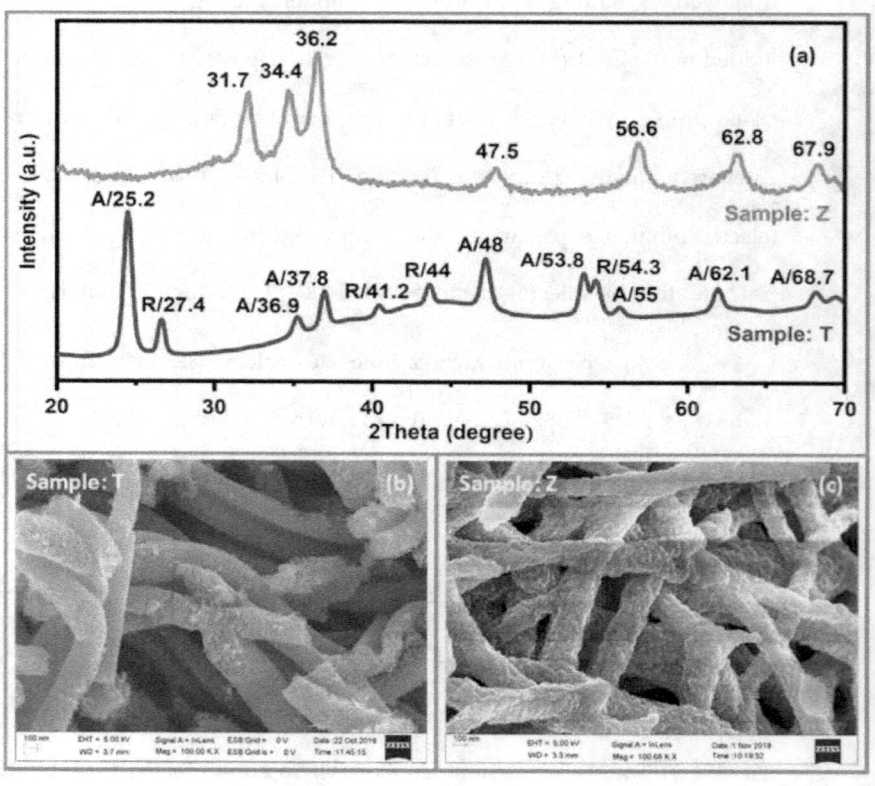

Figure 5.3.1.(a)XRD patterns and 5.3.1(b),(c),FESEM images of TiO_2 (T) and ZnO (Z) nanofibers.

The XRD pattern of the T-sample shows the presence of two crystallographic forms that are anatase (A) and rutile (R) indicating the polycrystalline nature. The anatase peaks originated at Bragg angle 2θ=25.2º,

36.9º, 37.8º, 48º, 53.8º, 55º, 62.1º and 68.7º were assigned to the planes (101), (103), (004), (200), (105), (213) and (116) respectively. Similarly, the rutile peaks originated at 2θ=27.4º, 41.2º, 44º and 54.3º were assigned to the planes(110), (111), (210) (211) respectively. The anatase and rutile peaks were found in matching with the JCPDS File No. 21-1272 and JCPDS File No. 21-1276 respectively. The XRD pattern of the Z-sample endorses the characteristic diffraction peaks of the ZnO crystallite originated at 2θ=31.7º, 34.4º, 36.2º, 47.5º, 56.6º, 62.8º and 67.9º corresponding to the planes (100), (002), (101), (102), (110), (103) and (112).

The XRD pattern indicates the formation of hexagonal wurtzite crystalline phase and coincides with the JCPDS File No. 36-1451. No other peaks were noticed in both the samples, T and Z endorsing the absence of impurities. The morphology of the TiO_2 and ZnO nanofibers were investigated by FESEM as shown in **figure 5.3.1(b)** and **5.3.1(c)**. The prepared nanofibers of the TiO_2 and ZnO were found randomly aligned. Comparatively, TiO_2 nanofibers possess a smoother surface than the ZnO nanofibers. Using ImageJ tool, the mean diameter of the T and Z samples were observed to be 169 and 209 nm respectively.

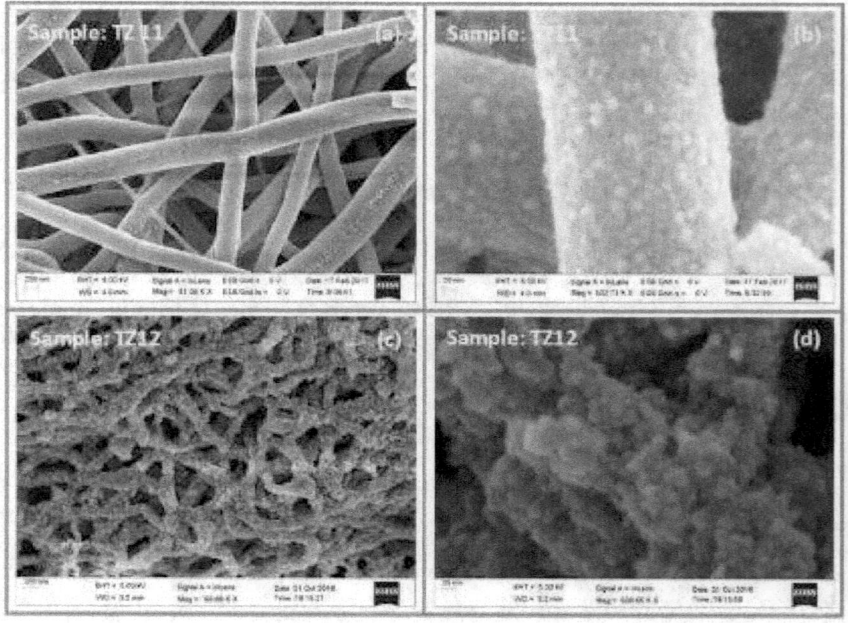

Figure 5.3.2(a),(b),(c),(d), **Surface morphology of TZ11 and TZ12 nanofibers.**

5.3.2. FESEM Analysis

Figure **5.3.2(a)** and **5.3.2(b)** depicts the FESEM images of the composite TiO_2/ZnO nanofibers (TZ11) prepared with the proportional ratio 1:1 of the TiO_2:ZnO solutions. Here, we can observe the preparation of the smooth and aligned nanofibers as depicted in **figure 5.3.2(a)**. A high scale FESEM image depicted in **figure 5.3.2(b)** shows the well-aligned nanofibers with the composition of TiO_2 and ZnO spherical nanoparticles in the sample TZ11.

Comparatively, TZ12 nanofibers were noticed to be thinner and non-uniform as depicted in **figure5.3. 2(c)**. However, **figure 5.3.2(d)** endorses the presence of TiO_2 and ZnO nanoparticles arranged in a random direction. The diameters of the TZ11 and TZ12 fibers were found to be in the range from 194-379 and 115-240 nm respectively. However, the average particles size was estimated to be 11 and 11.3 nm corresponding to the fiber samples TZ11 and TZ12. Here, the double proportion of the ZnO solution resulted in the rough fibers with the randomly arranged nanoparticles in the fiber form as referring to **figure 5.3.2(b)** and **5.3.2(d)**.

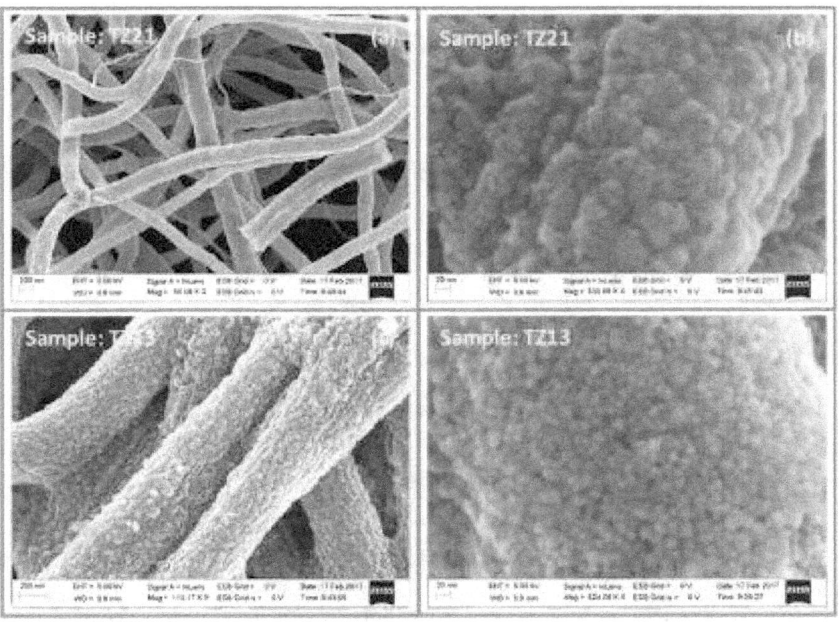

Figure 5.3.3(a),(b),(c),(d), Surface morphology of TZ21 and TZ13 nanofibers.

5.3.3. Surface morphology.

With the increased ZnO solution i.e. proportion T:Z=1:2, we have noticed rough nanofibers as depicted in **figure 5.3.2(c)** and 5.3.**2(d)**. Inversely with the sample TZ21, we can notice the preparation of the smoother nanofibers when the proportion of the TiO_2 solution was double than the ZnO i.e. T:Z=2:1 as shown in **figure 5.3.3(a)**. A closer look of the TZ21 fiber morphology depicted in **figure 5.3.3(b)** endorses the compact packing of the spherical nanoparticles. The diameter of the TZ21 and TZ13 fibers and nanoparticles were found to be in the range from 155-284 and 332-428 nm and 11 and 9.8 nm respectively. As depicted in **figure 5.3.3(c),** a distinct morphology of the TZ13 nanofibers can be noticed which is based on the proportional ratio, T:Z=1:3. With a similar trend as discussed in **figure5.3.2(c),** we can observe the formation of rough and well-aligned nanofibers. It means roughness of the fibers is enhanced with the increased concentration of the ZnO solution. However, well packing of the TiO_2 and ZnO nanoparticles with somewhat porosity is observed in the FESEM image as shown in **figure 5.3.3(d)**.

5.3.4. TEM Analysis.

With the interest of distinct morphology observed for the TZ13 sample, we have performed the transmission electron microscopy (TEM) investigation. **Figure 5.3.4 (a)** shows the cylindrical morphology of the TZ13 nanofibers with somewhat rough surface, as this was noticed in **figure 5.3.3(c)**. As compared to TZ11, TZ12 and TZ21 composite nanofibers, the diameter of the TZ13 sample is found to be increased in the range 184-464 nm with the increased proportion of the ZnO solution as discussed in **figure 5.3.2** and **5.3.3**. TEM result coincides with the FESEM observation and the reported work [22]. As depicted in **figure5.3.4(b),** the diameter of the nanofiber is found to be c.a. 230 nm which is made-up of grain-like nanocrystals of the TiO$_2$ and ZnO. **Figure 5.3.4(c)** depicts the high-resolution TEM (HR-TEM) image of the TZ13 nanofibers while the d-space imaging is shown in the inset of **figure 5.3.4(c)**. The lattice d-spacing is found to be 0.298 nm which corresponds to the (100) plane of the hexagonal wurtzite ZnO as discussed later. The selected area electron diffraction (SAED) image shown in **figure5.3.4(d)** endorses the polycrystalline nature of composite TZ13 nanofibers.

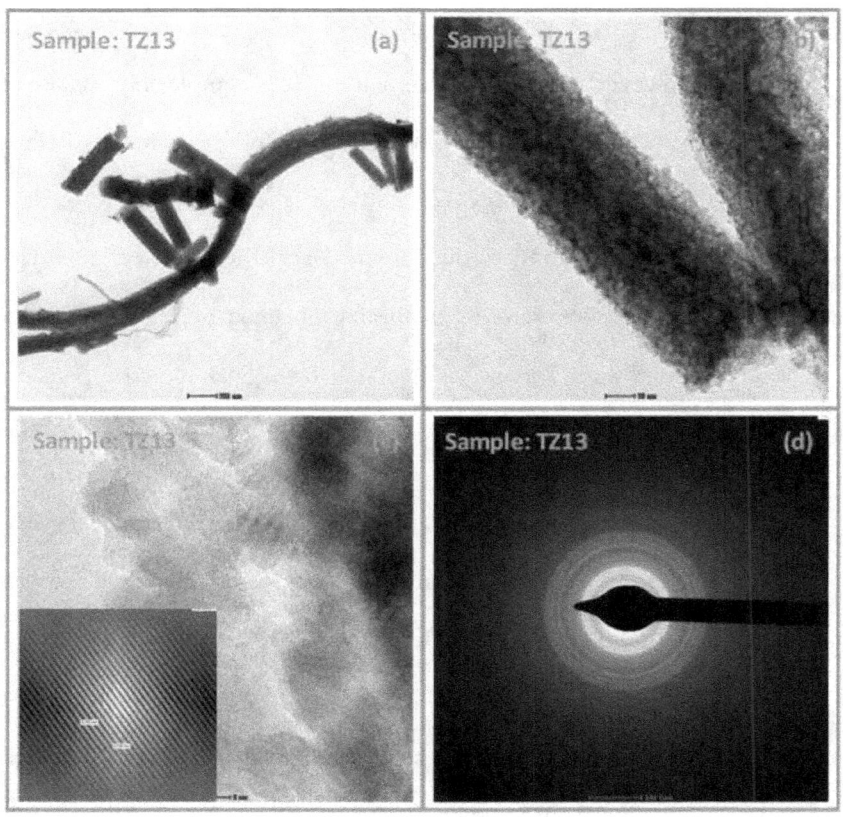

Figure 5.3.4. (a),(b),(c),(d), TEM micrographs of TZ13 nanofibers fig.(a) and fig.(b), HR-TEM fig.(c) and SAED pattern fig.(d).

To verify the crystallinity of the composite TZ13 nanofibers, XRD study was performed and the obtained pattern shows the good crystallinity as shown in **figure 5.3.5(a)**. The anatase TiO_2 peaks can be observed at $2\theta=25.2°$, $40.4°$, $48°$ and $53.8°$ while wurtzite ZnO peaks at $2\theta=31.7°$, $34.4°$, $36.2°$, $47.5°$, $56.6°$, $62.8°$ and $67.9°$ are also exhibited. A peak at $2\theta=22.5°$ can also be observed which may be associated with the polymer residual [21]. We can recall that the XRD pattern of the T-sample endorsed the mixed crystalline phases of the anatase and rutile as shown in **figure 5.3.1(a)**.

However, in the XRD pattern of the TZ13-sample only anatase phase of the TiO_2 and ZnO peaks are evidenced. More ZnO peaks in the XRD pattern is regarded the increased ZnO concentration as compared to TiO_2. Depending on the ZnO concentration, the dominant peaks may vary as reported in the literature [22]. Therefore, the d-spacing obtained by HR-TEM investigation represents the ZnO plane (100) of the hexagonal wurtzite polycrystalline phase.

Under UV illumination, the semiconductor metal-oxides like TiO_2, ZnO etc. produce oxidized hydroxyl and oxyradicals due to the generation of the electron-hole pairs and decomposes the organic materials into less dangerous contents. Using TZ11, TZ12, TZ21 and TZ13 composite fibers as the catalysts, we have studied the photodegradation of the Eriochrome black T (EBT) dye in aqueous under UV irradiation.

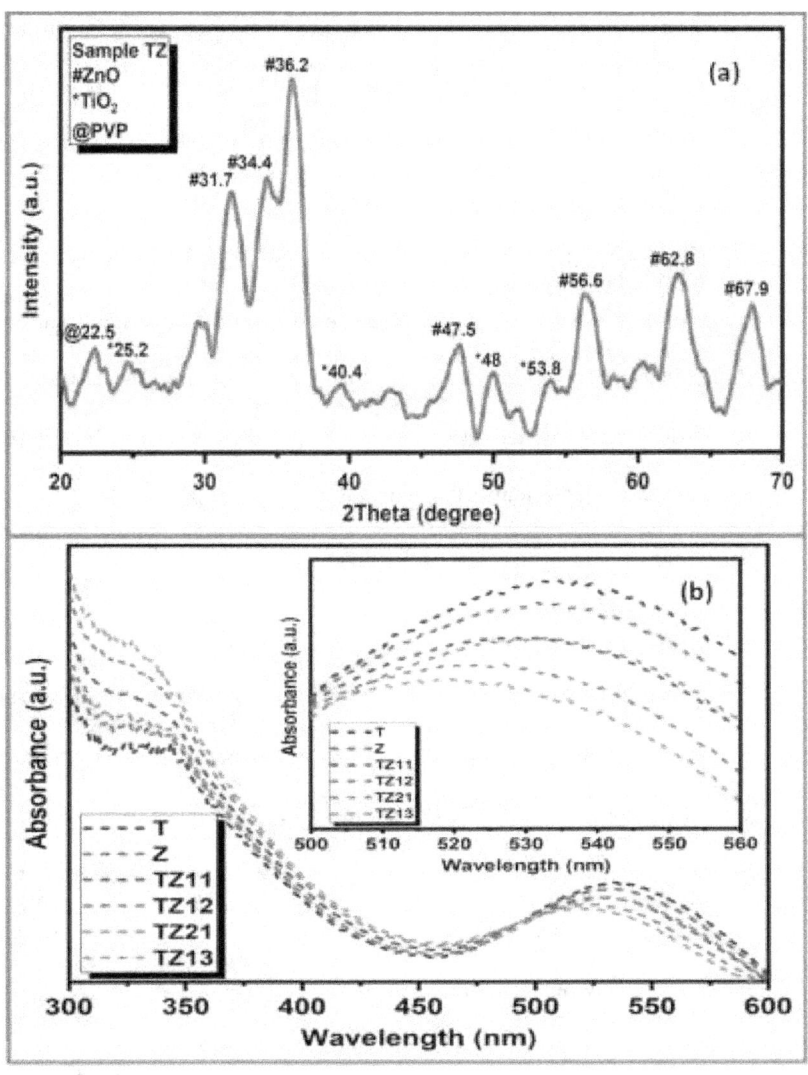

Figure 5.3.5(a),(b), XRD pattern of TZ13 nanofibers fig.(a) and Uv-vis absorbance of variant composite nanofibers fig.(b)

Figure 5.3.5(b) depicts the UV-vis absorbance recorded in a regular interval upto 120 min of UV irradiation. A significant photodegradation of the EBT dye can be observed when the proportion of the ZnO is increased.

Comparatively, TZ13-sample exhibits the best photodegradation among the TZ11, TZ12 and TZ21 samples. The anatase TiO$_2$ absorbs efficiently UV light and moreover, rutile phase does not exist in our case as depicted in **figure 5.3.5(a)**.Furthermore, the charge carriers recombination gets minimal due to the presence of the ZnO [22]. The improvement in the photocatalytic activity is ascribed to the synergetic effect of the TiO$_2$/ZnO composite nanofibers with their distinct morphology. The nanofibers possess a higher surface area than the nanoparticles which increases the active sites. As a result, this boosts the adsorption of the dye and therefore, the enhanced photocatalytic activity was exhibited [17].

5.4. Conclusions

We have investigated the structural and morphological properties of the various metal oxides (T, Z, and TZ) nanofibers. The XRD patterns of the pure-TiO$_2$ showed the mixed anatase and rutile phases while the hexagonal wurtzite phase was noticed of the ZnO. Remarkably, XRD pattern of the TZ13 composite nanofibers exhibited the anatase peaks corresponding to the TiO$_2$ along with the ZnO-wurtzite peaks.

The XRD pattern of the TZ13 composite nanofibers is dominated with the major peaks of the ZnO which could be regarded the increased proportion of the ZnO solution as compared to TiO$_2$. Morphology studied by FESEM evidenced the smooth, long and randomly aligned nanofibers for all the samples T, Z, TZ11, TZ12, TZ21 and TZ13. However, a distinct morphology of the TZ13 nanofibers is attributed to the increased proportion of the ZnO solution (T:Z=1:3). Though TZ13 nanofibers are found to be well-aligned with its c.a. diameter 230 nm but noticed to be rougher.

TEM investigation endorsed the well-aligned nanofibers with the lattice d-spacing around 0.298 nm which is regarded the (100) plane of the wurtzite hexagonal phase of the ZnO. The selected area electron diffraction (SAED) analysis showed the polycrystalline nature of the TZ13 nanofibers and coincided with the XRD result. Furthermore, the photodegradation of the EBT dye using TZ11, TZ12, TZ21 and TZ13 catalysts were performed whereas TZ13-sample endorsed the best photodegradation of the EBT dye.

References

1. Agnieszka Kołodziejczak-Radzimska and Teofil Jesionowski, Zinc Oxide-From Synthesis to Application: A Review, Materials (Basel)., Vol. 7(4), 2833–2881 (2014).

2. Adawiyah J.Haider, Zainab N.Jameel, Imad H.M.Al-Hussaini, Review on: Titanium Dioxide Applications, Energy Procedia Vol. 157, 17-29 (2019).

3. Xueyan Li, Desong Wang, Guoxiang Cheng, Qingzhi LuobJing An, Yanhong, Wang, Preparation of polyaniline-modified TiO_2 nanoparticles and their photocatalytic activity under visible light illumination, Applied Catalysis B: Environmental, Vol. 81, Issues 3–4, 267-273 (2008).

4. Navin Jain, Aprit Bhargava and Jitendra Panwar, Enhanced photocatalytic degradation of methylene blue using biologically synthesized "protein-capped" ZnO nanoparticles, Chemical Engineering Journal, Vol. 243, 549-555 (2014).

5. Jintao Tian, Lijuan Chen, Yansheng Yin, Xin Wang, Jinhui Dai, Zhibin Zhu, Xiaoyun Liu and Pingwei Wu, Photocatalyst of TiO_2/ZnO nano composite film: Preparation, characterization, and photodegradation activity of methyl orange, Surface & Coatings Technology 204 (2009) 205-214.

6. D.L.Liao, C.A.Badour and B.Q.Liao, Preparation of nanosized TiO_2/ZnO composite catalyst and its photocatalytic activity for degradation of methyl orange, Journal of Photochemistry and Photobiology A: Chemistry, 194/1, 11-19 (2008).

7. Carina Chun Pei and Wallace Woon-Fong Leung, Enhanced photocatalytic activity of electrospun TiO_2/ZnO nanofibers with optimal anatase/rutile ratio, Catalysis Communications 37 (2013) 100-104.

8. Adem Yar, Bircan Haspulat, Tugay Üstün, Volkan Eskizeybek, Ahmet Avcı, Handan Kamış and Slimane Achour, Electrospun TiO_2/ZnO/

PAN hybrid nanofiber membranes with effcient photocatalytic activity, RSC Adv., 7, 29806-29814 (2017,).

9. Jeong-Ha Baek, Juyun Park, Jisoo Kang, Don Kim, Sung-Wi Koh, and Yong-Cheol Kang, Fabrication and Thermal Oxidation of ZnO Nanofibers Prepared via Electrospinning Technique, Bull. Korean Chem. Soc., Vol. 33, No. 8 (2012) 2694.

10. Jian Li, Long Yan, Yufei Wang, Yuhong Kang, Chao Wang and Shaobo Yang, Fabrication of TiO_2/ZnO composite nanofibers with enhanced photocatalytic activity, J Mater Sci: Mater Electron (2016).

11. Evando S. Araújo, Bruna P. da Costa, Raquel A.P. Oliveira, Juliano Libardi, Pedro M. Faia and Helinando P. de Oliveira, TiO_2/ZnO hierarchical heteronanostructures: Synthesis, characterization and application as photocatalysts, Journal of Environmental Chemical Engineering 4 (2016) 2820-2829.

12. A.F. Lotus, S.N. Tacastacas, M.J. Pinti, L.A. Britton, N. Stojilovic, R.D. Ramsier and G.G. Chase, Fabrication and characterization of TiO_2–ZnO composite nanofibers, Physica E 43 (2011) 857-861

13. Carina Chun Pei and Wallace Woon-Fong Leung, Photocatalytic degradation of Rhodamine B by TiO_2/ZnO nanofibers under visible-light irradiation, Separation and Purification Technology 114 (2013) 108-116.

14. Xian Liu, Yan-yu Hu, Ri-Yao Chen, Zhen Chen, and Hong-Chun Han, Coaxial Nanofibers of ZnO-TiO_2 Heterojunction With High Photocatalytic Activity by Electrospinning Technique, Synthesis and Reactivity in Inorganic, Metal-Organic, and Nano-Metal Chemistry, 44:449-453, 2014.

15. Delong Li, Xudong Jiang, Yupeng Zhang, and Bin Zhang, A novel route to ZnO/TiO_2 heterojunction composite fibers, J. Mater. Res., Vol. 28, No. 3, Feb 14, 2013.

16. Sun Hye Hwang, Jooyoung Song, Yujung Jung, O. Young Kweon, Hee Song and Jyongsik JangElectrospun ZnO/TiO_2 composite nanofibers as a bactericidal agentw, Chem. Commun., 2011, 47, 9164–9166.

17. Jia Dong Chen, Wei Sha Liao, Ying Jiang, Dan Ni Yu, Mei Ling Zou, Han Zhu, Ming Zhang and Ming Liang Du, Facile Fabrication of ZnO/TiO_2 Heterogeneous Nanofibres and Their Photocatalytic Behaviour and Mechanism towards Rhodamine B, Nanomater Nanotechnol, 2016, 6:9, doi: 10.5772/62291.

18. Muzafar A. Kanjwal, Nasser A. M. Barakat, Faheem A. Sheikh, Soo Jin Park and Hak Yong Kim, Photocatalytic Activity of $ZnO-TiO_2$ Hierarchical Nanostructure Prepared by Combined Electrospinning and Hydrothermal Techniques Macromolecular Research, Vol. 18, No. 3, pp 233-240 (2010).

19. Ruilai Liu, Huiyan Ye, Xiaopeng Xiong and Haiqing Liu, Fabrication of TiO_2/ZnO composite nanofibers by electrospinning and their photocatalytic property, Materials Chemistry and Physics 121 (2010) 432-439.

20. Jin-Ah Park, Jaehyun Moon, Su-Jae Lee, Sang-Chul Lim and Taehyoung Zyung, Fabrication and characterization of ZnO nanofibers by electrospinning, Current Applied Physics 9 (2009) S210–S212.

21. M. H. Abou_Taleb, Thermal and Spectroscopic Studies of Poly(N-vinyl pyrrolidone)/Poly(vinyl alcohol) Blend Films. Journal of Applied Polymer Science, Vol. 114, 1202–1207 (2009).

22. Jia Dong Chen, Wei Sha Liao, Ying Jiang, Dan Ni Yu, Mei Ling Zou, Ming Zhang, Ming Liang Du and Han Zhu, Facile Fabrication of ZnO/TiO_2 Heterogeneous Nanofibres and Their Photocatalytic Behaviour and Mechanism towards Rhodamine B, Nanomater Nanotechnol, Vol., 6/9, pp8, doi: 10.5772/62291, (2016).

Chapter-6

Conclusions and Future Remarks

6.1 Conclusions

An electrospinning approach is a versatile and simple technique for the fabrication of nanofibers for their potential applications. The prepared samples were studied by x-ray diffraction (XRD) and endorsed the mixed phases of anatase and rutile-TiO_2. TG/DTA study of TiO_2/PVP mat demonstrated the thermal response with respect to increased temperature. By performing the FTIR study, the presence of various functional groups along with the vibration peak corresponding to the Ti-O-Ti bond is found at 660 cm^{-1}. Before the optimization of process parameters, FESEM analysis revealed the formation of randomly distributed nanofibers with their diameter in the range from 244-343 nm. An increased applied voltage and distance from tip-collector respectively produced thinner nanofibers in the range from 293 nm-175 nm and 259 nm-147 nm.

Further, the formation of thin nanofibers of diameter 214 nm-111 nm and 284 nm-102 nm are observed with the decreased flow rate of solution and PVP concentration respectively. Finally, the optimized values of process parameters of electrospinning could decrease the diameter of TiO_2 nanofibers to 74 nm. DSSC photoanode was fabricated using thinner TiO_2 nanofibers, and the photovoltaic performance was evaluated. This process parameter study is beneficial to tailor diameter of the nanofibers either thin or thick according to application.

Furthermore, the work was extended to fabricate ZnO (Z) and TiO$_2$/ZnO (TZ) composite nanofibers along with TiO2 (T). The XRD patterns of pure-TiO$_2$ (T) revealed the mixed anatase and rutile phases while the ZnO (Z) sample evidenced the hexagonal wurtzite structure. To study the morphology of the TiO$_2$/ZnO (TZ) composite nanofibers, the concentration of TiO$_2$ and ZnO solutions were varied. Significantly, the TZ13 composite nanofibers endorsed the TiO$_2$-anatase peaks along with the ZnO-wurtzite peaks as revealed by the XRD pattern. In the case of TZ13 composite nanofibers, XRD pattern is found to be dominated by the ZnO peaks as compared to TiO$_2$ which is attributed to the increased concentration of ZnO solution.

FESEM Morphology endorsed that the samples T, Z, TZ11, TZ12, TZ21, and TZ13 were smooth, long, and randomly aligned nanofibers. Nonetheless, a distinct morphology of TZ13 nanofibers is due to the increased ZnO solution (T: Z=1:3). TEM investigation evidenced the formation of well-aligned nanofibers with the lattice d-spacing about 0.298 nm, which is designated the (100) ZnO hexagonal-wurtzite plane. The study of selected area electron diffraction (SAED) revealed the polycrystalline nature of TZ13 nanofibers and found in good agreement with the XRD pattern. In addition, photodegradation was carried out for the EBT dye using catalysts TZ11, TZ12, TZ21, and TZ13, whereas the TZ13 sample endorsed the best photodegradation for the EBT dye.

6.2 Future Remarks

Based on its distinct physical and chemical properties of 1D nanostructure 'nanofibers' have their potential demand in the context of scientific, medical, and industrial applications. Nanofiber is a type of nanomaterial whose diameter can be scaled from tens to hundreds of nanometer which makes these materials unique due to its large surface-to-volume ratio. Nanofibers have the ability to produce the extremely porous mesh which could be periodically fabricated with air voids for their amazing applications. The substantial impact of nanofibers technology can indeed be ascribed by the variety of basic materials that can be used for the fabrication of nanofibers.

In other words, fundamental materials like natural polymers, synthetic polymers, carbon substances, semiconductor substances and hybrid materials can be used to realize the advanced applications. The present scenario of nanofibers demand has been extended in commercial products like fabrics, filtering, wound/cut repairing, human body organs like knee implantation and many more. Nanofibers are being recognized as promising materials dye-sensitized solar cells, energy storage, batteries, pollutant/water treatment, environmental control, medical surgery, etc.

In this context, our work is useful to understand the tuning of physical, chemical, mechanical, thermal and biological properties by considering the electrospinning parameters in order to achieve the better performance of the advanced devices based on either single „TiO$_2$' or composite „TiO$_2$/ZnO' nanofibers. This thesis is having the scope for the new comer scholar where he/she can use this work for the further application point of view.

LIST OF PUBLICATIONS

1) Electrospinning process parameters dependent investigation of TiO_2 nanofibers
   M.V. Someswararao, R.S. Dubey, P.S.V. Subba Rao and Shyam Singh
   Results in Physics, 11, 223-231, 2018.

2) Experimental Investigation of Electrospun Titania Nanofibers: An Applied Voltage Influence
   M.V. Someswararao, D. Pradeep, R.S. Dubey and P.S.V. Subba Rao
   Materials Today: Proceedings 18, 384–388, 2019.

3) Fabrication and characterisation of electrospun barium titanate and polyvinly pyridine composite nanofibers
   Adavi Bala Krishna, **M.V. Someswararao**, P.S.V. Subba Rao, R.S. Dubey, B.S. Diwakar and K.S. Jaideep
   Materials Today: Proceedings, 18, 2142-2146, 2019.

4) Preparation and Investigation of TiO2/ZnO Composite Nanofibers for Photocatalytic Application
   M.V.Someswararao, R.S. Dubey, P.S.V. Subba Rao and
   Prof. K.V. Ramesh
   Communicated to Materials Today Communications (2020)

Electrospinning process parameters dependent investigation of TiO_2 nanofibers

M.V. Someswararao[a], R.S. Dubey[b,*], P.S.V. Subbarao[c], Shyam Singh[d]

[a] *Department of Physics, SRKR Engineering College, Bhimavaram, A.P., India*
[b] *Department of Nanotechnology, Swarnandhra College of Engineering and Technology, Seetharamapuram, Narsapur, A.P., India*
[c] *Department of Physics, Andhra University, Visakhapatnam, A.P., India*
[d] *Department of Physics, University of Namibia, Windhoek, Namibia*

ARTICLE INFO

Keywords:
Electrospinning
Nanofibers
Process parameters
X-ray diffraction
Surface morphology

ABSTRACT

This paper reports the electrospinning fabrication and characterization of TiO_2 nanofibers. Various electrospinning process parameters such as applied voltage, distance tip-collector, solution flow rate and polymer (PVP) concentration are studied. The prepared nanofibers were investigated by X-ray diffraction (XRD), scanning electron microscopy (SEM), energy dispersive X-ray spectroscopy (EDS), thermogravimetric-differential thermal analysis (TG-DTA) and Fourier transform infrared spectroscopy (FTIR). XRD pattern of TiO_2 nanofibers evidenced the presence of mixed phases of anatase and rutile. TG/DTA investigation showed the characteristic peaks corresponding to the heating behavior of TiO_2-PVP mat. FTIR investigation endorsed a vibration peak at $660\,cm^{-1}$ associated with the characteristic Ti–O–Ti bond. With the optimized process parameters, TiO_2 nanofibers diameter was found to be reduced to 74 nm as compared to the first sample prepared with the diameter of 343 nm. Furthermore, these nanofibers were employed as the photoanode material for the preparation of dye sensitized solar cell (DSSC) and photovoltaic study is evaluated.

Introduction

Titanium dioxide (TiO_2) is a versatile material which has been well-recognized for its several applications such as cosmetics, protective surface coatings, solar cells, sensors (including chemical, gas & bio), water treatment, paints, batteries and many more. The key demand of this material is its non-toxicity, high chemical stability, bio-degradability and low-cost production. There are three main crystallite phases of TiO_2 exists such as anatase and rutile in tetragonal while the brookite in orthorhombic shapes. Out of these phases, anatase and brookite can transform to rutile phase via heat treatment whereas rutile remains stable. Sol-gel-derived TiO_2 possesses the anatase phase however; other phases can be attained by controlling the heat treatment mechanism or by preferring the more acidic solution during the synthesis. The phase transformation could be understood by the two mechanisms; surface energy and precursor chemistry. For the case of anatase phase, the associated surface energy is weaker as compared to others two phases. Further, the geometry of the crystal structure is governed by the precursor chemistry which involves the nucleation and the growth of either phases [1].

One-dimensional (1D) TiO_2 nanostructures such as nanotubes, nanowires, nanobelts, nanofibers, etc. have been demanded owing to their fast electron-transport and carrier-collection capability. TiO_2 nanofibers are being investigated as photoanode material in dye-sensitized solar cells (DSSCs) application. However, photoanode based on 1D-TiO_2 nanostructures possess low efficiency mainly due to the weak dye-adsorption associated with the surface morphology. Conversely, by treating the surface with acid, $TiCl_4$ and oxygen plasma one can improve the dye loading. DSSCs based on TiO_2 nanofibers were studied to investigate the influence of the surface treatment [2]. The improved conversion efficiency was noticed in the order 8.59% < 9.33% in accordance with the photoanodes treated with the acid and oxygen plasma. The morphology of TiO_2 nanofibers showed the great influence with the calcination temperature. The photocatalytic activity of TiO_2 nanofibers calcined at 500 °C for 3 h was performed using rhodamine under visible light irradiation and found satisfactory as compared to nanofibers calcined at temperature 600, 700 °C for 3 h [3]. Electrospun TiO_2 nanofibers in the range from 194 to 441 nm have been investigated for the application as the scattering material for the DSSC [4]. The scattering property was found to be linearly dependent on the diameter and density of the fibers. The photocurrent-voltage characteristics of the DSSCs were evidenced the increased performance

https://doi.org/10.1016/j.rinp.2018.08.054
Received 10 July 2018; Received in revised form 28 August 2018; Accepted 30 August 2018
Available online 05 September 2018
2211-3797/ © 2018 The Authors. Published by Elsevier B.V. This is an open access article under the CC BY-NC-ND license
(http://creativecommons.org/licenses/BY-NC-ND/4.0/).

which has been attributed to the scattering effect caused by TiO_2 nanofibers. Further, the photocatalytic activity of hydrogen evolution under UV irradiation was studied which endorsed the similar scattering effect as compared to other samples. Various morphologies of electrospun TiO_2 nanofibers have been investigated for DSSCs and photocatalytic applications. By co-axial electrospinning, the hollow/tubular TiO_2 nanofibers were prepared and further etching treatment was preferred using sodium hydroxide aqueous solution in order to get the porous morphology of TiO_2 nanofibers [5]. The diameter of the hollow/tubular nanofibers was in the range of 300–500 nm whereas porous nanofibers were in ribbon shape with their width about 200 nm. Brunauer-Emmett-Teller (BET) surface area of the hollow/tubular TiO_2 nanofibers was 27.3 m^2/g, which was almost double the solid TiO_2 nanofibers; however 106.5 m^2/g surface area was obtained for the porous TiO_2 nanofibers. The morphological study of TiO_2 nanofibers via solution viscosity and electrospinning process parameters has been investigated [6]. The low viscous solution with the high ratio of precursor solution and glacial acetic acid resulted in the beaded morphology of TiO_2 nanofibers. The optimized process parameters showed the smooth nanofibers with their average diameters 148 ± 79 nm. Mesoporous TiO_2 nanofibers prepared by electrospinning process followed by the solvothermal treatment has also been studied for their photocatalytic activity [7]. The additional solvothermal process was found useful to crystallize TiO_2 by arranging closely packed grains which could improve the adsorption of CO_2. As a consequence, enhanced photocatalytic behavior of TiO_2 was noticed which has been regarded as the improved adsorption and the charge separation influenced by the solvothermal process. The conductivity of TiO_2 nanofibers are low owing to its higher resistivity. Therefore, doped-TiO_2 nanofibers have been investigated for the enhancement of conductivity. In addition, enhanced conductivity of TiO_2 nanofibers have been reported by treating with potassium hydroxide which could alter the insulating behavior into conductivity due to their reduced resistivity [8]. This behavioral change of TiO_2 makes this material apposite for supercapacitor application. Potassium hydroxide treated TiO_2 showed the abrupt increase in the magnitude of the capacitance value which was about 1500 times the pristine. Gold doped-TiO_2 nanofibers were prepared and their photocatalytic activity was studied [9]. An enhanced photocatalytic activity was observed with gold-doped TiO_2 nanofibers. This result was attributed to the formation of the Schottky-barrier at the junction of gold-TiO_2 which prevented the carriers recombination and hot electron generation. In similar way, electrospun silver-TiO_2 nanofibers were reported to study the influence of silver concentration [10]. With the increased doping concentration of silver, the diameter of fibers was found to be increased while photoluminescence intensity was weaker. The antibacterial study was performed with the pathogenic bacteria and found improved as compared to pure-TiO_2 nanofibers. The enhanced antibacterial activity has been attributed to the silver doping along with the large surface area of the prepared nanofibers. In another approach, graphitic carbon nitride nanosheets hybridized nitrogen-doped titanium dioxide nanofibers in-situ fabrication has been reported by electrospinning process [11]. The prepared hybrid nanofibers were mesoporous structure and its partial decomposition was found responsible for the doping of nitrogen in the bulk-TiO_2. As maximum as 931.3 $\mu molh^{-1} g^{-1}$ photocatalytic H_2 production rate was achieved which has been associated with the graphitic-C_3N_4 nanosheet hybridized N-doped TiO_2 nanofibers. This hybrid material was found promising for the improved light absorption and the electron-transport mechanism. Electrospun silver/TiO_2 nanofibers were studied for the photocatalytic activity prepared at different sintering temperatures and silver concentrations [12]. The nanofibers calcined at temperature 500 °C was about 120 nm in diameter with 20 nm their particles size which showed about 71% degradation rate of methylene blue. The porous and uniform TiO_2/g-C_3N_4 (graphitic-carbon nitride composite) nanofibers prepared by electrospinning method have been reported [13]. The diameter was found to be 100–150 nm after calcination of nanofibers at temperature 550 °C. To study the photocatalytic activity of the prepared nanofibers, the degradation of rhodamine B dye under sunlight was evaluated. The photocatalytic activity was found to be increased which has been associated with the hetero-junction TiO_2/g-C_3N_4 and was promising for the charge transport mechanism along with the prevention of charge recombination. Over the randomly aligned nanofibers, unidirectional nanofibers possess enhanced optical and mechanical properties with their high degree of crystallinity. These unidirectional grown nanofibers favor the better transport of charge-carriers and therefore, enhanced the performance of the devices [14,15]. A study of electrospun TiO_2 nanofibers by employing the modified aluminum collector of two-pieces has been reported which endorsed the unidirectional growth of nanofibers [16]. Further, this study was explored for the tuning of nanofibers diameter by controlling the tip-collector distance and the applied voltage.

Various techniques such as drawing, template synthesis, electrospinning, self-assembly and phase separation and are available for the preparation of polymer-based nanofibers. Among the aforementioned techniques, an electrospinning fabrication technique is recognized to be promising for the growth of continuous fibers due to its easy and cost-effective process. Electrospinning system consists of three main parts; (1) metal collector (drum/plate/disk, etc.), (2) syringe pump with metal tip/needle and, (3) dc high voltage power supply. The morphology of the electrospun fibers is significantly governed by the process parameters. These process parameters are the applied dc voltage, solution/gel flow rate, distance metal tip-collector and polymer concentration. The diameter of nanofibers decrease with the increase of applied dc voltage and the distance tip-collector. However, these conditions are valid for enough viscous solution otherwise it produces beads/particles by electrospraying rather than electrospinning process. In a similar way, the reduced solution flow rate and the polymer concentration yields thinner nanofibers. In addition, the ambient environment like humidity and temperature have their significant role in the preparation of continuous nanofibers without any defect.

In this work, we present the optimization of electrospinning process parameters for the preparation of TiO_2 nanofibers. The diameter of the nanofibers showed the great influence of the applied voltage, the distance tip-collector, the solution flow rate and the polymer (PVP) concentration. Section 'Experimental details' describes the experimental approach for the preparation- of TiO_2 nanofibers by electrospinning process. The characterized results have been discussed in Section 'Results and discussion'. Finally, Section 'Conclusions' presents the summary of the paper.

Experimental details

Materials

For the preparation of TiO_2 nanofibers, titanium tetraisopropoxide (TTIP, Sigma-Aldrich), acetic acid solution (Sdfine), polyvinyl pyrrolidone (PVP, Mw = 1,300,000, Sigma-Aldrich) and methanol (Fisher Scientific) were procured and used without any further purification.

Electrospinning setup

The electrospinning setup is illustrated in Fig. 1. It mainly consists of collector drum, dc power supply and syringe pump. These all mechanisms are assembled in a fume hood whereas its front panel shows the various controls like the speed of collector-drum, applied dc voltage, spin rate and flow rate. The right-hand side top image depicts the enlarged image of the collector drum while fiber Taylor cone formation can be observed in the bottom image.

Methods

Before the electrospinning process, the sol-gel synthesis was

Fig. 1. Electrospinning setup for the preparation of nanofibers.

performed to get enough viscous solution. At first, 0.6 ml TTIP precursor was vigorously stirred in 10 ml methanol. After 5 min, 4 ml glacial acetic acid was added in TTIP solution and kept for 30 min stirring at room temperature. Later, 1.12 g PVP was dissolved in the above solution and stirred for 3 h. The prepared solution was found transparent and enough viscous which was loaded into a syringe. For the first electrospinning process, the flow rate and the distance tip-collector were fixed to 3 ml/h and 10 cm respectively whereas the applied voltage was maintained to 12 kV. The images of peeling off the as-prepared TiO$_2$-PVP mat and the collected one are shown in Fig. 2(a) and (b) respectively. Later, optimization of the process parameters was performed by varying the applied voltage, distance tip-collector, flow rate and the polymer concentration.

Characterization

The electrospun TiO$_2$ nanofibers were characterized to examine the phase and crystallinity using X-ray Diffraction (XRD, Bruker AXS D8 Advance, Germany), the qualitative and quantitative analysis using Fourier-transform infrared spectroscopy (FTIR, Shimadzu, Japan), heating behavior of TiO$_2$-PVP mat using thermogravimetric-differential thermal analysis (TG-DTA, DTG-60H, Shimadzu), surface morphology study using scanning electron microscope (SEM, JSM-6360, USA) and the chemical composition investigation using EDS attached to SEM.

Results and discussion

X-ray diffraction (XRD) measurement was carried out to investigate the crystalline nature of TiO$_2$ nanofibers. Fig. 3 depicts the XRD patterns of TiO$_2$-PVP mat and TiO$_2$ nanofibers calcined at 450 °C for 3 h.

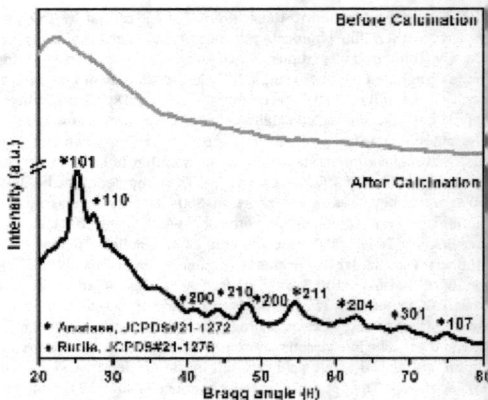

Fig. 3. X-ray diffraction patterns of electrospun TiO$_2$-PVP mat and calcined TiO$_2$ nanofibers.

Before calcination, no appearance of the diffraction peak indicates the amorphous nature of TiO$_2$-PVP mat. Inversely, various characteristic diffraction peaks can be observed for the case of calcined TiO$_2$ nanofibers at 450 °C for 3 h. The XRD pattern indicates the mixed anatase and rutile phases which were assigned to JCPDS#21-1272 and JCPDS#21-1276 respectively. The highest intensity diffraction peak at $2\theta = 25$ of the plane (1 0 1) corresponds to the anatase phase of TiO$_2$. The anatase peaks were originated from the lattice planes at 2θ values $25\theta = d_{101}$, $48\theta = d_{200}$, $54\theta = d_{211}$, $62\theta = d_{204}$ and $74\theta = d_{107}$ while rutile peaks at $27\theta = d_{110}$, $41\theta = d_{200}$, $44\theta = d_{210}$ and $69\theta = d_{301}$.

The Scherrer's formula was employed to estimate the crystallite size of the calcined TiO$_2$ nanofibers. The Scherrer's equation is represented as $d = k\lambda/\beta\cos\theta$, where d is the crystallite size, $k = 0.89$ is constant dependent on the crystalline shape, λ is the X-ray wavelength 1.54056 Å for Cu Kα, β is the full width at half-maximum intensity, and θ is the Bragg angle. The estimated crystallite size was found to be 8.2 nm corresponds to the most predominant diffraction peak (1 0 1) of anatase phase.

Fig. 4 depicts the surface morphology of TiO$_2$ nanofibers calcined 450 °C for 3 h. We can observe the randomly aligned and smooth morphology of TiO$_2$ nanofibers at scale 1 μm as shown in Fig. 4(a) and at scale 200 nm in Fig. 4(b). The diameter of TiO$_2$ nanofibers was found

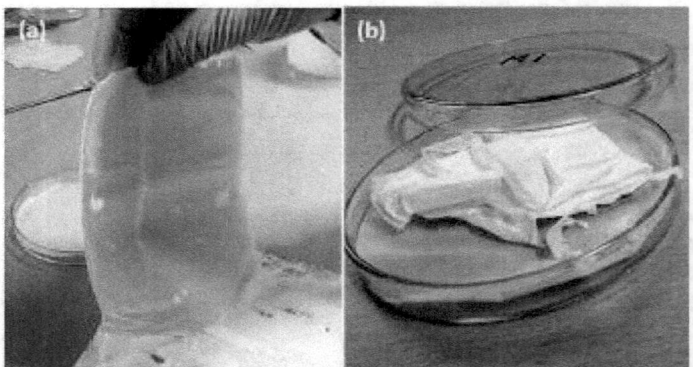

Fig. 2. Peeling off the as-prepared electrospun TiO$_2$-PVP mat on aluminum foil figure (a) and collected sample figure (b).

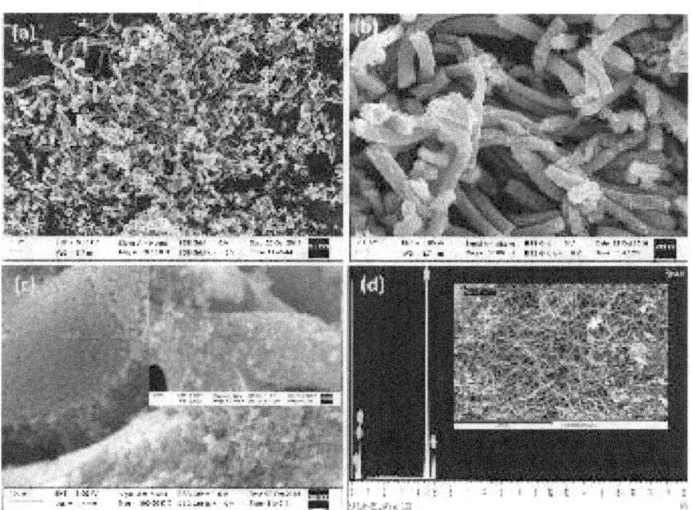

Fig. 4. Surface morphology of calcined TiO$_2$ nanofibers at scale 1 μm figure (a), 200 nm figure (b), 100 nm figure (c) and EDS spectra figure (d).

be in the range of 244–343 nm. Fig. 4(c) endorses the TiO$_2$ particulate at 100 nm scale while the inset image was recorded at 20 nm scale. TiO$_2$ nanoparticles diameter was found to be 11 nm as estimated using the jImage open source software. The elemental composition of Ti and O were also confirmed by EDS measurement as shown in Fig. 4(d).

To investigate the compositional changes during thermal treatment, thermogravimetric-differential thermal analysis (TG/DTA) of TiO$_2$-PVP composite nanofibers was carried out. As depicted in Fig. 5, the TGA curve endorses the three stages of weight loss. The first region 0–100 °C corresponds to evaporation of residual solvents including desorption of water content in the sample. The next weight loss is up to 76% in the temperature region from 100 to 300 °C indicating the removal of polymer contents from TiO$_2$-PVP mat. However, the third steep weight loss is observed in the region 300–550 °C which reveals the maximum degradation of the polymer with its loss up to 12%. Referring to DTA curve, an endothermic peak aligned at 80 °C is assigned to the evaporation of moisture and the solvent while an exothermic peak at 390 °C can be observed which is attributed to the decompositions of metal hydroxide and polymer. The peak at 500 °C represents the phase transformation from anatase to rutile [17–20]. However, no mass loss was observed after 640 °C.

This work is limited to the analyses of optical and structural properties of TiO$_2$ nanofibers nonetheless the mechanical property is another significant parameter particularly, when the fibers are subjected to mechanical stress during its usage for example as the water filter or so. In brief, the as-prepared TiO$_2$ nanofibers were studied by employing the cantilevered beam bending approach attached to scanning electron microscope [21]. The electrospun nanofibers prepared with and without needle have been studied for Young's modulus and bending strength analyses. By investigations, the electrospun TiO$_2$ nanofibers prepared with needle showed the uniform morphology with their better mechanical property. In similar way, the mechanical property for the application of nanofiber-reinforced polymer composites has been investigated [22]. For the sample under the test, the hooking and elongation were precisely controlled and the proposed approach was found suitable regardless the both end gripping of nanofibers. The nanofibers based on various diameters have been characterized however, the small diameter based nanofibers showed the better mechanical strength and therefore, it was suggested as the nano-reinforcement for the composite materials.

Further, we have prepared various samples of TiO$_2$ nanofibers for the optimization of process parameters such as the applied voltage, the distance tip-collector, the flow rate and the PVP or polymer concentration. Accordingly, Figs. 6–9 depicts the SEM images of TiO$_2$ nanofibers and their analyses are presented in the further discussion.

Fig. 6(a), (b), (c) and (d) depicts the morphology of electrospun TiO$_2$ nanofibers prepared at voltages 8, 9, 10 and 11 kV respectively. We can observe the smooth and randomly distributed nanofibers with their estimated diameters 293, 226, 189 and 175 nm in accordance with the applied voltages 8, 9, 10 and 11 kV. The distance from tip-collector, flow rate and the PVP concentration were kept at 10 cm, 1 ml/h and 1 g respectively. As the applied voltage was increased the diameter of nanofibers was found to be reduced from 293 nm to 175 nm. Here, we can understand that an optimal applied voltage is important to evaporate the solvent faster while stretching the fiber towards the collector.

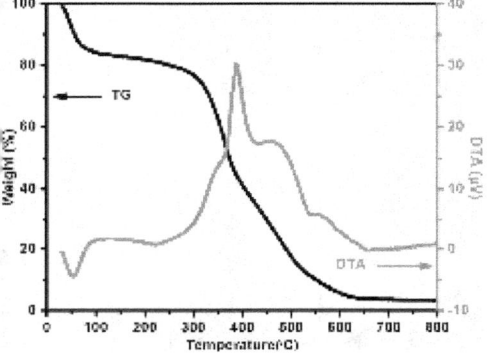

Fig. 5. TG/DTA graph of electrospun TiO$_2$-PVP mat.

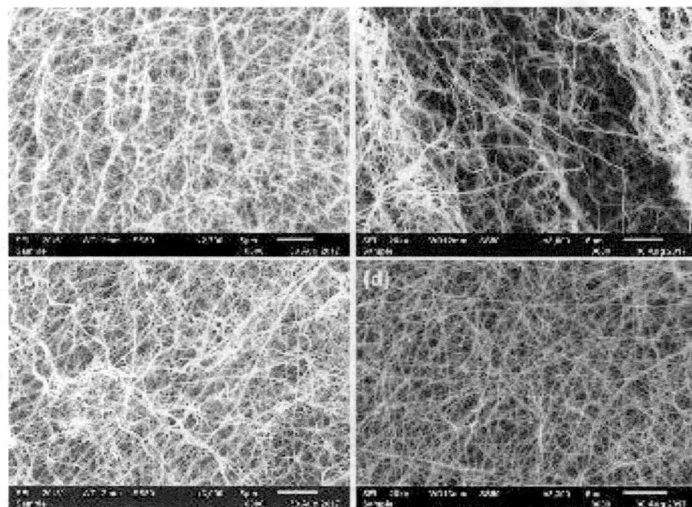

Fig. 6. SEM images of TiO$_2$ nanofibers prepared at voltages 8, 9, 10 and 11 kV.

The surface morphology of electrospun TiO$_2$ nanofibers prepared at various tip-collector distances 8, 9, 12 and 14 cm are illustrated in Fig. 7(a), (b), (c) and (d) respectively. Here, the parametrical values of the applied voltage, flow rate and the PVP concentration were 10 kV, 1 ml/h and 1 g respectively. The average diameters were found to be 259, 189, 167 and 147 nm corresponding to the tip-collector distance 8, 9, 12 and 14 cm respectively. This analysis reveals that the distance from the tip-collector is a prominent parameter which directly affects the evaporation time required for the solvent and as a result, the fast evaporation process leads to thinner fibers.

We can observe the formation of randomly distributed smooth nanofibers in Fig. 8 as a function of solution flow rate. The average diameters were found to be 111, 155, 189 and 247 nm in accordance with the flow rates of 0.6, 0.8, 1.0 and 1.2 ml/h. During this optimization process, the applied voltage, tip-collector distance and the PVP concentration were maintained to 10 kV, 10 cm and 1 g respectively. Here, the reduced diameter (111 nm) of nanofibers shows the importance of an optimum flow rate of the solution to get the thinner fibers without any beads.

Finally, PVP concentration was optimized while keeping the applied voltage, tip-collector distance and flow rate to 10 kV, 10 cm and 1 ml/ respectively. The SEM images of TiO$_2$ nanofibers prepared wi

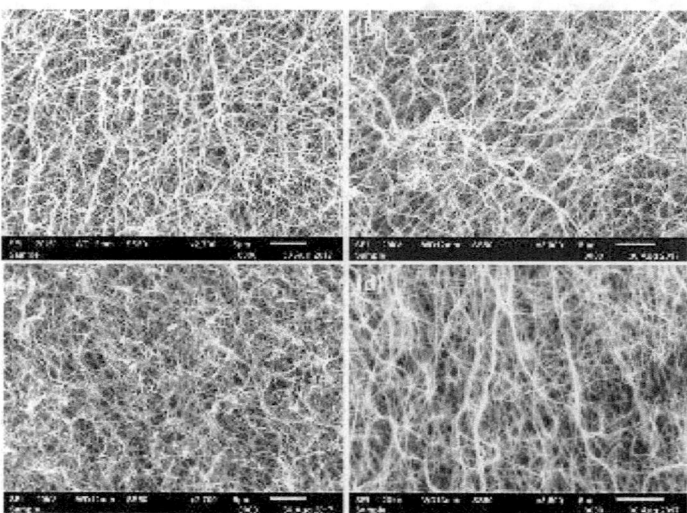

Fig. 7. SEM images of TiO$_2$ nanofibers prepared at distances 8, 10, 12 and 14 cm.

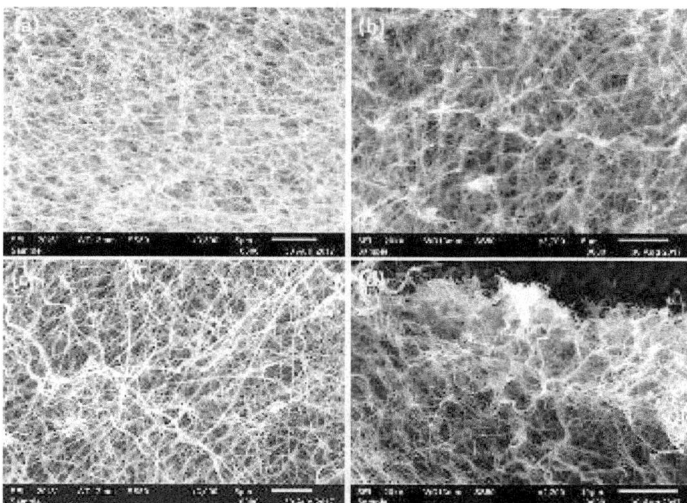

Fig. 8. SEM images of TiO$_2$ nanofibers prepared at flow rates 0.6, 0.8, 1.0 and 1.2 ml/h.

different PVP concentrations are shown in Fig. 9. The diameter of the nanofibers was found to be 102, 152, 189 and 284 nm with respect to the PVP concentrations 0.6, 0.8, 1.0 and 1.2 g. Depending upon the optimal PVP concentration, thin nanofibers can be obtained as it is observed in Fig. 9(a). The influence of the four process parameters such as the applied voltage, distance tip-collector, flow rate and the PVP concentration is summarized in Fig. 10.

In general, the applied voltage must satisfy the minimum required voltage that is exceeding the threshold voltage for the ejection of charged jets from the Taylor cone so that the electrospinning process gets started. An increased voltage enhances the electrostatic force on the solution which causes the stretching of jet and hence, leads to thin nanofibers. Accordingly, a decrease in nanofibers diameter from 293 to 175 nm can be noticed in Fig. 10(a). The morphology of the nanofibers has great influence of the applied voltage as observed in SEM images shown in Fig. 6. In brief, an optimum voltage is recommended to have thin fibers whereas higher voltage results in beads formation. The distance from tip-collector is another parameter which influences the surface morphology and diameter of the nanofibers as observed in Fig. 7. Therefore, a minimum distance is required for the evaporation of the solvent that too before the fiber reaches to the collector during the electrospinning process. As the distance of tip-collector increases the

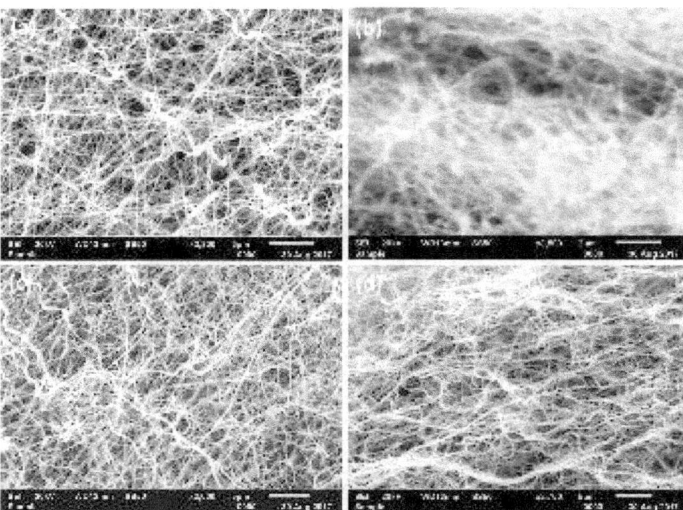

Fig. 9. SEM images of TiO$_2$ nanofibers prepared at PVP concentrations 0.6, 0.8, 1.0 and 1.2 g.

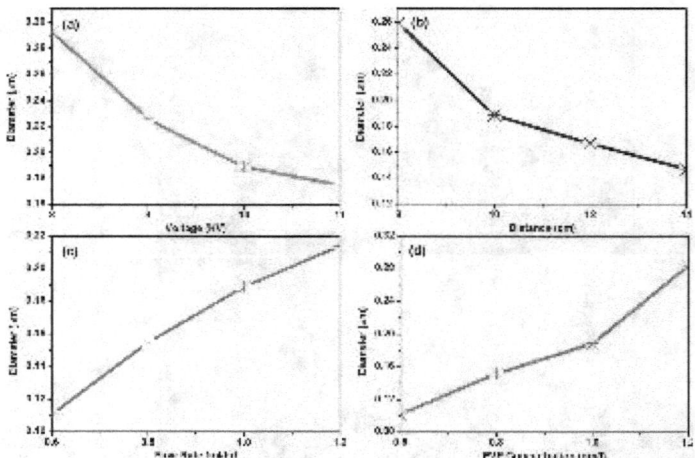

Fig. 10. Diameter of TiO$_2$ nanofibers as a function of applied voltage, distance tip-collector, flow rate and the PVP concentration.

diameter of the fibers reduces from 259 to 147 nm as shown in Fig. 10(b). A large distance produces the thin fibers however; beads formation can be prevented by avoiding too near or too far distance from the tip-collector. The flow rate is a significant parameter which deals with the polymer solution transfer rate and its speed. Generally, sufficient time is needed for the solvent evaporation which can be attained by choosing a small flow rate. An optimal flow rate is fine for the smooth fiber preparation however, a high flow rate yields beaded fibers as it gets lesser time for the solvent evaporation. Fig. 10(c) depicts the increment in fibers diameter from 111 to 214 nm as a function of flow rate. In general, the high flow rate decreases the charge density which produces the fibers with a large-diameter. It means an increase in feed rate yields a corresponding increase in the diameter of the fibers. At last, PVP concentration is another important parameter which yields thinner fibers from 284 to 102 nm by decreasing the concentration of the polymer as evidenced in Fig. 10(d). An optimum concentration of polymer yields smoother and thinner fibers without any beads as observed in Fig. 9. In this way, the diameter of the electrospun nanofibers can be tuned by optimizing the applied voltage, the distance tip-collector, the flow rate and the polymer concentration. An electrospinning approach is recognized as an easy and low-cost process whereas the reproducibility of the nanofibers can be attained by opting the optimal values of these parameters.

After optimizing the process parameters, we have obtained the optimal values of applied voltage 11 kV, distance tip-collector 14 cm, the flow rate of 0.6 ml/h and PVP concentration of 0.6 g. Finally, we have performed an electrospinning process while maintaining the above parametrical values. Fig. 11 shows the SEM image and EDS spectra of TiO$_2$ nanofibers prepared by keeping the optimal values of the process parameters. We can observe the continuous and randomly oriented nanofibers with their average diameter of 74 nm in Fig. 11(a). The reduced diameter of the nanofibers is attributed to the optimized parameters. The elemental composition of Ti and O are evidenced in the EDS spectra as depicted in Fig. 11(b).

Further, the analysis of chemical bonding and compositions were performed by FTIR measurement in the range of 4000–400 cm^{-1} as shown in Fig. 12(a). The various vibration peaks were observed and found good in agreement with reported literature [17–19]. Accordingly, the peak at 3432 cm^{-1} represents the O–H stretching vibration associated with the absorbed water. An asymmetric peak at 2930 cm^{-1} represents the stretching vibration of C–H group while another peak at 2850 cm^{-1} is attributed to the symmetric stretching mode of C–H. The stretching mode of C=O vibration is observed at 1640 cm^{-1} whereas peak at 1457 cm^{-1} endorses the bending vibration of CH$_2$ group. The C–N group of polymer associated with the asymmetric stretching vibration is assigned at 1113 cm^{-1} and the vibration peak at 660 cm^{-1} is found associated with the characteristic Ti–O–Ti bonds.

This work is limited to the investigation of electrospun TiO$_2$ nanofibers via various process parameters however, we have extended the testing of TiO$_2$ nanofibers as the photoanode material of the dye-sensitized solar cell (DSSC). In brief, 0.15 g of TiO$_2$ nanofibers of diameter 74 nm was mixed in 0.25 ml ethanol, 0.25 ml acetic acid and 0.25 ml of

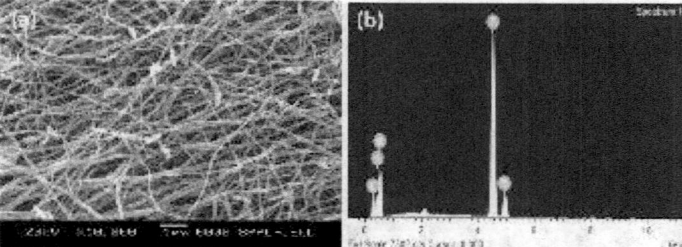

Fig. 11. SEM image figure (a) and EDS spectra figure (b) of TiO$_2$ nanofibers prepared after the optimization of electrospinning process parameters.

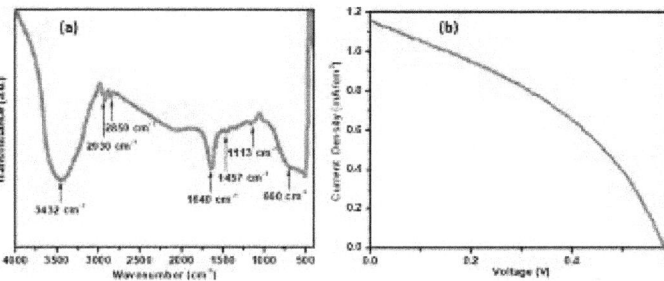

Fig. 12. FTIR spectra of electrospun TiO$_2$ nanofibers figure (a) and current density-voltage characteristics of DSSC figure (b).

VP to get slurry. The fluorine-doped (FTO) glasses of 1 cm^2 were used the electrodes after ultrasonically cleaned in ethanol and acetone multaneously. By doctor blade method, TiO$_2$ paste was rolled-on FTO ass within the active area 0.25 cm^2. After coating, it was dried at a mperature 60 °C and finally sintered at 500 °C for 30 min. Later, the otoanode was soaked in rhodamine B dye for 12 h and then used after ashing with ethanol and de-ionized water. For the preparation of a unter electrode, the platinum paste (Plastisol T, Solaronix) was rolled the FTO glass using doctor blade method. The assembly of DSSC was ne by clipping the counter electrode on top of the dye loaded photoanode with offsetting the electrodes in the opposite direction for the ligator wires connections. To prevent the shorting between the unter-electrode and the photoanode, an insulating film was placed hich was kept open at one offset edge for the insertion of the electrolyte solution. After dropping the electrolyte solution, the binder clips re slightly opened and closed in order to allow the solution in the tive area of the cell. For the measurement of the current density-ltage characteristic of DSSC, Keithley 2420 power source and white D source with 80 mW/cm^2 illumination was employed. Fig. 12(b) ows the current density-voltage characteristic of DSSC which shows s conversion efficiency 0.38% with open circuit voltage 0.58 V, short-rcuit current 1.15 mA/cm^2 and 0.39 fill factor. DSSC performance can e enhanced by using thinner nanofibers as the host material with timizing the preparation of photoanode [23]. Besides, photoanode sed on the hybrid structure of TiO$_2$ nanoparticles and nanofibers as e scattering have been reported which showed the enhanced photo-ltaic performance [24,25].

nclusions

The electrospinning fabrication and characterization of TiO$_2$ nano-ers have been reported. XRD measurement endorsed the mixed atase and rutile phases. TG/DTA investigation evidenced the char-teristic peaks of TiO$_2$/PVP mat. By FTIR study, a vibration peak at 0 cm^{-1} corresponding to characteristic Ti–O–Ti bond is observed. M measurement showed the randomly distributed nanofibers with eir diameter in the range of 244–343 nm before the optimization. rthermore, optimization of various electrospinning process para-eters was performed. An increased applied voltage and the distance -collector led to the preparation of thinner nanofibers from 3–175 nm and 259–147 nm respectively. While the decreased flow te and PVP concentration evidenced the further thin nanofibers from 4–111 nm and 284–102 nm respectively. Furthermore, the optimized lues of electrospinning process parameters could reduce the diameter TiO$_2$ nanofibers to 74 nm. Using thinner TiO$_2$ nanofibers, DSSC otoanode was fabricated and photovoltaic performance was eval-ted. Finally, this study is helpful to optimize the electrospinning ocess parameters for the preparation of thin/thick nanofibers and eir reproducibility using an easy and in expensive method.

Acknowledgement

RSD acknowledges partial help from Mr. D. Pradeep (JRF) for the sol-gel synthesis.

References

[1] Reyes-Coronado D, Rodríguez-Gattorno G, Espinosa-Pesqueira ME, Cab C, RdeCoss, Oskam G. Phase-pure TiO2 nanoparticles: anatase, brookite and rutile. Nanotechnology 2008;19:145605. 10pp.
[2] Li Y, Ding JN, Yuan NY, Bai L, Hu HW, Wang XQ. The influence of surface treatment on dye-sensitized solar cells based on TiO2 nanofibers. Mater Lett 2013;97:74–7.
[3] Li Jing, Qiao Hui, Du Yuanzhi, Chen Chen, Li Xiaolin, Cui Jing, et al. Electrospinning synthesis and photocatalytic activity of mesoporous TiO2 nanofibers. Sci World J. 2012. https://doi.org/10.1100/2012/154939. 7 pages. Article ID 154939.
[4] Chen Yuan-Lian, Chang Yi-Hao, Huang Jow-Lay, Chen Ingann, Kuo Changshu. Light scattering and enhanced photoactivities of electrospun titania nanofibers. J Phys Chem C 2012;2012(116):3857–65.
[5] He Guangfei, Cai Yibing, Zhao Yong, Wang Xiaoxu, Lai Chuilin, Xi Min, et al. Electrospun anatase-phase TiO2 nanofibers with different morphological structures and specific surface areas. J Colloid Interface Sci 2013;398:103–11.
[6] Sadeghi Soraya Mirmohammad, Vaezi Mohammadreza, Kazemzadeh Asghar, Jamjah Roghayeh. Morphology enhancement of TiO2/PVP composite nanofibers based on solution viscosity and processing parameters of electrospinning method. J Appl Polym Sci 2018;135:46337. 11pg.
[7] Junwei Fu, Cao Shaowen, Yu Jiaguo, Low Jingxiang, Lei Yongpeng. Enhanced photocatalytic CO-reduction activity of electrospun mesoporous TiO2 nanofibers by solvothermal treatment. Dalton Trans 2014;43:9158–65.
[8] He X, Yang CP, Zhang GL, Shi DW, Huazg QA, Xiao HB, et al. Supercapacitor of TiO2 nanofibers by electrospinning and KOH treatment. Mater Des 2016;106:74–80.
[9] Wang Tao, Zhang Yu, Wang Yong, Wei Jinxin, Zhou Ming, Zhang Zhengmei, et al. One-step electrospinning method to prepare gold decorated on TiO2 nanofibers with enhanced photocatalytic activity. J Nanosci Nanotechnol 2018;18:3176–84.
[10] Kudhier MA, Sabry RS, Al-Haidarie YK, AL-Marjani MF. Significantly enhanced antibacterial activity of Ag-doped TiO2 nanofibers synthesized by electrospinning. Mater Technol 2017. https://doi.org/10.1080/10667857.2017.13967.
[11] Han Cheng, Wang Yingde, Lei Yongpeng, Wang Bing, Wu Nan, Shi Qi, et al. In situ synthesis of graphitic-C3N4 nanosheet hybridized N-doped TiO2 nanofibers for efficient photocatalytic H2 production and degradation. Nano Res 2015;8:1199–209.
[12] Dong Hong Ying, Sun Qing Hong, Zhang Ting Ting, Ren Qi, Ma Wen. Synthesis and photocatalytic activity of ag doped TiO2 nanofibers. Mater Sci Forum 2018;913:1027–32.
[13] Tang Qian, Menga Xianfeng, Wang Zhiying, Zhou Jianwei, Tang Hua. One-step electrospinning synthesis of TiO2/g-C3N4 nanofibers with enhanced photocatalytic properties. Appl Surface Sci 2018;430:253–62.
[14] Jalili R, Morshed M, Ravandi SAH. Fundamental parameters affecting electrospinning of PAN nanofibers as uniaxially aligned fibers. J Appl Polym Sci 2006;101:4350–7.
[15] Kovtyukhova NI, Mallouk TE. Nanowires as building blocks for self-assembling logic and memory circuits. Chem Eur J 2002;8:4354–63.
[16] Kim Jae-Hun, Lee Jae-Hyoung, Kim Jin-Young, Kim Sang Sub. Synthesis of aligned TiO nanofibers using electrospinning. Appl Sci 2018;8(309):10.
[17] Chang Meiqi, Sheng Ye, Song Yanhua, Zheng Keyan, Zhou Xiuqing, Zou Haiferg. Luminescence properties and Judd-Ofelt analysis of TiO2: Eu3+ nanofibers via polymer-based electrospinning method. RSC Adv 2016;6:52113–21.
[18] Chun Ho-Hwan, Jo Wan-Kuen. Polymer material-supported titania nanofibers with different polyvinylpyrrolidone to TiO2 ratios for degradation of vaporous trichloroethylene. J Ind Eng Chem 2014;20:1010–5.
[19] Chang Wenkai, Xu Fujian, Mu Xueyan, Ji Lili, Ma Guiping, Nie Jun. Fabrication of

nanostructured hollow TiO2 nanofibers with enhanced photocatalytic activity by coaxial electrospinning. Mater Res Bull 2013;48:2661–8.
[20] Nuansing Wiwat, Ninmuang Siayasunee, Jarernboon Wirat, Maensiri Santi, Seraphin Supapan. Structural characterization and morphology of electrospun TiO2 nanofibers. Mater Sci Eng B 2006;131:147–55.
[21] Vahtrus Mikk, Sutka Andris, Vlassov Sergei, Sutka Anna, Polyakov Boris, Dorogin Leonid, et al. Mechanical characterization of nanofibers using a nanomanipulator and atomic force microscope cantilever in a scanning electron microscope. Mater Charact 2015;100:98–103.
[22] Hwang Kenny Yoonki, Kim Sung-Dae, Kim Young-Woon, Woong-Ryeol Yu. Mechanical characterization of nanofibers using a nanomanipulator and atomic force microscope cantilever in a scanning electron microscope. Polym Test 2010;29:375–80.
[23] Jinchu I, Sreekala CO, Sajeev US, Achuthan K, Sreelatha KS. Photoanode engineering using TiO2 nanofibers for enhancing the photovoltaic parameters of natural dye sensitised solar cells. J Nano- and Electron Phys 2015;7:04002. 4pp.
[24] Lee Ji-Hye, Ahn Kyun, Kim Soo Hyung, Kim Jong Man, Jeong Se-Young, Jin Jor Sung, et al. Thickness effect of the TiO2 nanofiber scattering layer on the performance of the TiO2 nanoparticle/TiO2 nanofiber-structured dye-sensitized solar cells. Curr Appl Phys 2014;14:856–61.
[25] Anjusree GS, Deepak TG, Narendra Pai KR, Joseph John, Arun TA, Nair Shantikumar V, et al. TiO2 nanoparticles @ TiO2 nanofibers – an innovative one dimensional material for dye-sensitized solar cells. RSC Adv 2014;4:22941–5.

ICAMME-2018

Experimental Investigation of Electrospun Titania Nanofibers: An Applied Voltage Influence

M.V. Someswararao[a], D. Pradeep[b], R.S. Dubey[b*], P.S.V. Subbarao[c]

[a]*Department of Physics, SRKR Engineering College, Bhimavaram, (A.P.), India*

[a]*Advanced Research Laboratory for Nanomaterials and Devices,* [b]*Department of Nanotechnology, Swarnandhra College of Engineering and Technology,Seetharamapuram, Narsapur, (A.P.), India.*

[c]*School Department of Physics, Andhra University, Visakhapatnam (A.P.), India*

Abstract

Electrospun nanofibers are being demanded in energy applications because of their easy and in-expensive fabrication process. In this paper, we present the fabrication of nanofibers at various different voltages 8, 9, 10 and 11kV. The nanofibers fabricated at 1kV could yield diameter 175 nm. Scanning electron microscopy (SEM) investigations confirmed the smooth and randomly aligned nanofibers whereas EDX analysis showed the elemental peaks of Ti and O. X-ray diffraction study revealed the mixed phases of anatase and rutile and FTIR measurement endorsed the stretching peak of Ti-O in the range from 800-500 cm^{-1}.

© 2019 Elsevier Ltd. All rights reserved.
Selection and/or Peer-review under responsibility of International Conference on Advances in Materials and Manufacturing Engineering, ICAMME-2018.

Keywords: TiO2 nanofibers; electrospinning; anatase and Brookite phases; Surface Morphology;

1. Introduction

Titanium dioxide (TiO$_2$) nanofibers have got demanded in several applications such as batteries, sensors, dye-sensitized solar cells, membranes, drug delivery etc. Presently, dye-sensitized solar cells (DSSCs) have received the attention of the scientific community due to their good energy conversion efficiency and easy fabrication process. DSSCs are expected as the promising candidate over the conventional silicon-based solar cells due to their low-cost and environmental friendly. After the pioneer work by the Gratzel, an intensive research efforts have been paid towards the high efficiency DSSCs and ~13 % has been reported so far [1,2]. A dye-sensitized solar cell with high permeability is a promising candidate as the third generation solar cell with the 20 % fabrication cost of the silicon based solar cells. The major components of the DSSCs are the nanocrystalline TiO$_2$ coated fluorine-doped tin-oxide glass as the photoanode, a dye as the sensitizer, a platinum coated fluorine-doped tin-oxide glass as the counter electrode and an electrolyte solution. In DSSCs, the role of dye is to adsorb on the photoanode which yields ejection of electrons towards the conduction band of the TiO$_2$ upon irradiance.

Instead of other factors affecting the DSSCs performance, the dye adsorption to the semiconductor oxide is very important while the nanocrystalline particles have limit due to its less surface area. Therefore, one-dimension nanostructure (nanotubes/nanofibers/nanowires) based photoanodes are suitable due to their improved surface area in addition to the direct-electron movement path while sustain the sufficient dye adsorption via percolating into the porous nanostructure. Further, the photoanode made of uniform TiO_2 layer with high surface area can improve the photovoltaic performance. Several papers have been reported on semiconducting metal-oxide such as titanium oxide, tin oxide and zinc oxide [3,4]. Among these, titanium dioxide is widely used material due to its photo activity, bio-compatibility, chemical and thermal stability, lightweight and low cost. Various fabrication methods are investigated to fabricate the TiO_2 nanostructures such as the hydrothermal method, template growth, thermal evaporation and electrospinning [5-7]. In spite these all, electrospinning method is noticed to be easy and inexpensive in process which yields continuous fibers of the desired diameter from some micron-nanometer size. By tuning the fabrication parameters such as dc voltage, solution flow rate, polymer concentration and distance between the jet and the collector plate, one can optimize the surface morphology of the fibers. During electrospinning fabrication, the polymer based precursor solution is ejected via a metal nozzle in the surrounding area of high dc field. Due to applied field, the electrostatic charges get build-up which evaporates solvent and therefore, the solution is stretched to obtain continuous fibers. In this way, the continuous flow of the solution give fibers which are finally, collected on the collecting drum/plate and later, the heat treatment is provided to crystalize the amorphous product required to get the host material based fibers after by eliminating the polymer contents.

In this paper, we study the influence of applied voltage on the diameter of TiO_2 nanofibers prepared by an electrospinning process. The experimental details of the preparation of fibers are presented in Section 2. The experimental results are discussed in Section 3. Finally, Section 4 concludes the paper.

2. Experimental Details

The following chemical are used as procured without any further purification. Titanium tetraisopropoxide, TIP Ti(OCH(CH3)2)4, (Sigma-Aldrich), Acetic Acid solution (Sdfine), poly (vinyl pyrolidone) (PVP) (C6H9NO)n with Mw=1,300,000 (Sigma-Aldrich) and Methanol.

For the sol-gel process, 0.5 ml TTIP precursor solution was mixed in, 10 ml methanol under stirring and later 4 ml acetic acid solution was added under vigorous stirring for 30min at room temperature. To this solution, 1g PVP was mixed and the stirring was continued up to 3 hr. The obtained solution was transparent and in gel-form. The as-prepared solution was loaded into the syringe pump having its needle 0.5 mm diameter which was vertically assembled. The flow rate and the collecting drum distance were maintained 1 ml/h and 10 cm respectively while applied voltage was varied as 8, 9, 10 and 11 kV.

The collected TiO2 fibers were characterized to examine the phase and crystallinity using X-ray Diffraction (XRD-Bruker AXS D8 Advance, Germany), the qualitative and quantitative analysis Fourier-transform infrared spectroscopy (FTIR-Shimadzu, Japan) , the surface morphology using scanning electron microscope (SEM, JSM-6360, USA) and the compositional chemical elementals investigation using EDX attached to SEM.

3. Results and Discussion

The surface morphology investigations of the samples prepared at dc voltage 8, 9. 10 and 11kV were carried using scanning electron microscopy, which are depicted in Fig. 1.

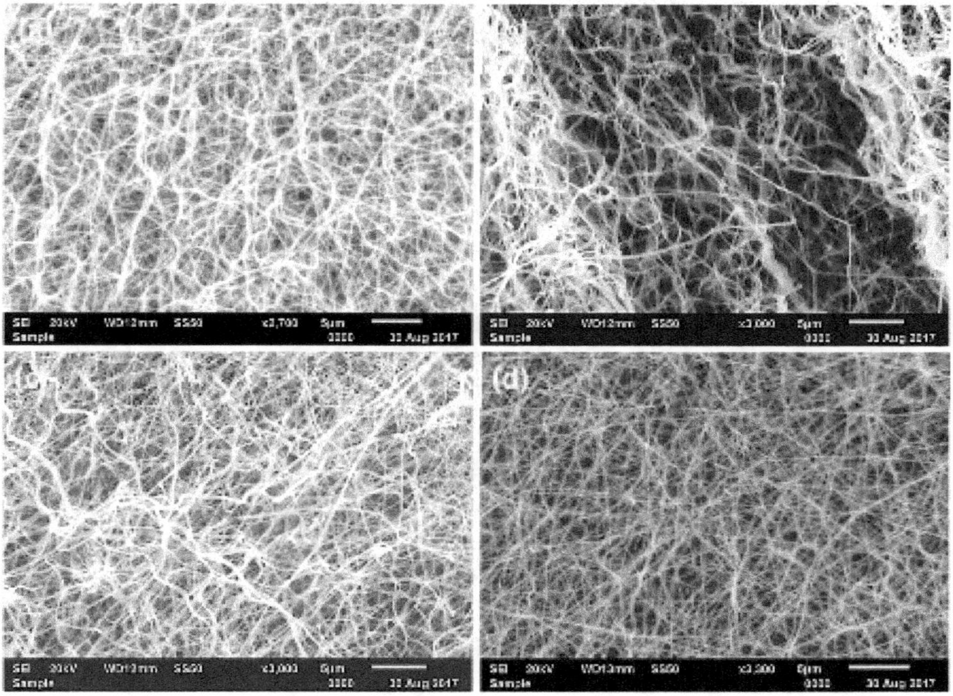

Fig. 1. SEM images of TiO$_2$ nanofibers prepared at voltages (a) 8kV, (b) 9kV, (c) 10kV, (d) 11kV calcined at temperature 600 0C for one hour.

The prepared nanofibers were found to be distributed in the random directions as can be observed in SEM images of Fig. 1(a), 1(b), 1(c) and 1(d) prepared at the voltages 8, 9, 10 and 11 kV respectively. The fibers are smooth and straight in morphology. Fig. 1(d) shows the thinner nanofibers with their average diameter ~175 nm electrospun at 11 kV dc voltage.

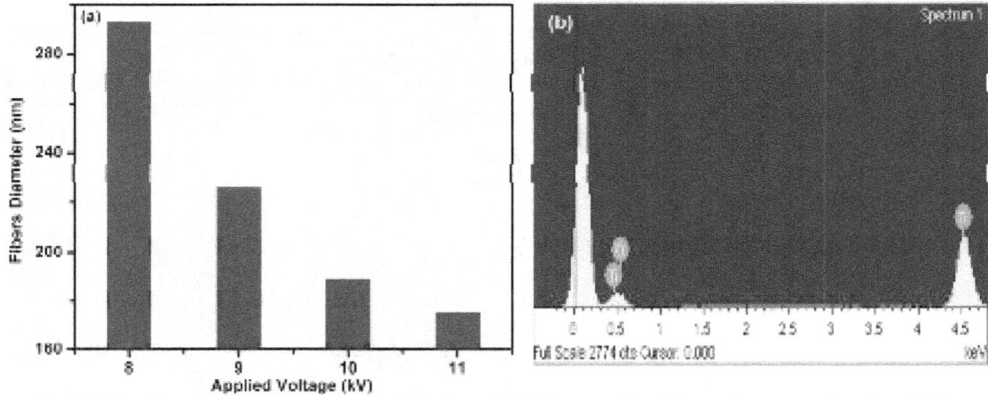

Fig. 2. (a) Nanofibers diameter as a function of applied voltage, (b) EDX spectra of the TiO$_2$ nanofibers prepared at voltage 11kV.

The analysis of nanofibers diameter in accordance to the applied voltage is presented in Fig. 2(a). The applied voltage is a significant parameter which directly affects the size of the nanofibers. As the dc voltage increases the fiber's diameter shrinks due to the fast evaporation of the solvent and the stretching of the fibers from the jet. The estimated sizes of the nanofibers were 293, 226, 189 and 175 nm corresponds to the applied voltages 8, 9, 10 and 11kV. Hereafter, the investigations were carried out for the thinner nanofibers shown in Fig. 1(d). Fig. 2(b) depicts the EDX spectra of the nanofibers which endorses the expected chemical elemental peaks of Ti and O.

The sample prepared with thinner nanofibers was investigated for the qualitative and quantitative analyses using FTIR measurement. Fig. 3 depicts the FTIR spectrum of the nanofibers with their estimated diameter 175 nm after the calcination at 600 °C for 1 hour.

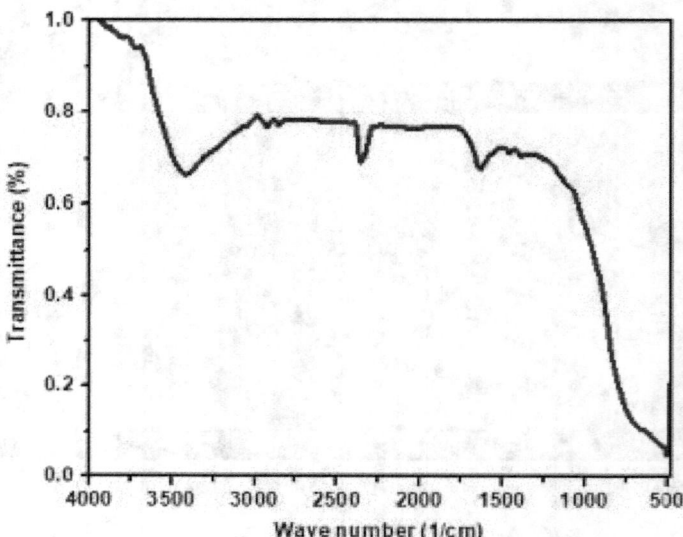

Fig. 3. FTIR spectrum of the calcined TiO_2 nanofibers.

FTIR measurement was carried out to the qualitative and quantitative analyses of the fiber sample calcined at 600^0C. A broad band can be noticed in the region 3416 cm^{-1} which is associated to the O-H stretching mode while C-H vibration peak can be observed at 2927 cm^{-1}. A peak at 2358 cm^{-1} corresponds to the C=C stretching mode which is associated to the alkyne group. A vibration peak of C=O stretching is assigned at 1632 cm^{-1} shows the C=O stretching however, a small peak at 504 $^{-1}$ is attributed to the Ti-O bond.

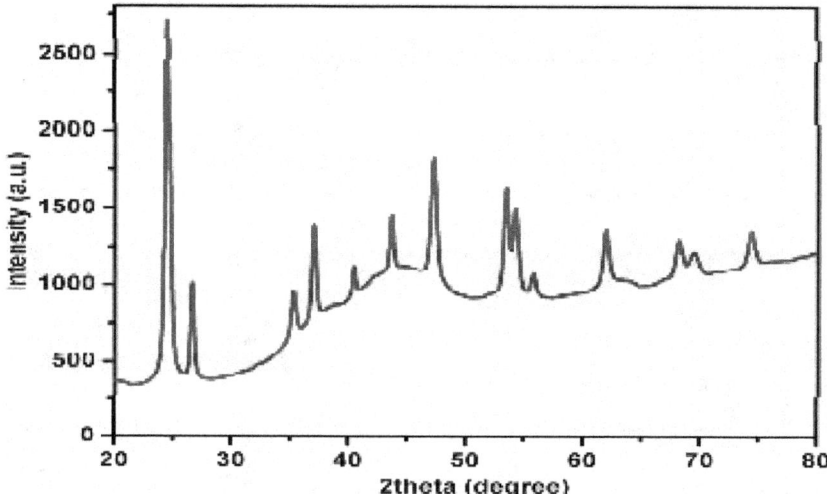

Fig. 4. XRD pattern of the calcined TiO$_2$ nanofibers.

To investigate the crystallinity and phase determination presented in the sample, XRD measurement was carried out. Fig. 4 depicts the XRD pattern of the nanofibers with the estimated size of 175 nm calcined for 1 hour at 600 °C. The XRD pattern shows the mixed phases of the anatase and brookite and attributed to the JCPDS file No, 83-224 and 86-1157 respectively. The diffraction peaks of anatase phase at 2θ values 25°, 36°, 53°and 74°can be ascribed to (101), (103), (105) and (107) planes respectively. Similarly,brookite phase at 2θ values 40°, 43°, 47°, 55°, 61°and 68°are corresponds to the (202), (412), (321), (230), (331) and (431) planes.

4. Conclusions

We have fabricated the TiO$_2$ nanofibers using an electrospinning unit and performed the various investigations. By varying the dc voltage, we have found the diameter of the nanofibers decreases in accordance to the increase of the applied voltage it. At 11kV voltage, the diameter is found to be 175 nm whereas 200 nm was attained at 8kV. The surface morphology of the nanofibers for the voltage variants was smooth and randomly aligned. Using EDX investigation, the compositional chemical elements Ti and O were found in the prepared sample. The XRD result showed the presence of mixed phases of the anatase and brookite. FTIR study showed the quantitative and qualitative analyses of the TiO$_2$ nanofibers and found satisfactory.

References

[1] M. Gratzel, J. Photochem. Photobiol. A, 164 (2004) 3-14.
[2] Simon Mathew, AswaniYella, PengGao, Robin Humphry-Baker, Basile F. E. Curchod, NegarAshari-Astani, IvanoTavernelli, Ursula Rothlisberger, Md. KhajaNazeeruddin and Michael Gratzel, Nature Chemistry 6 (2014)242–247.
[3] V. Baglio, M. Girolamo, V. Antonucci and A. S. Aricò, Int. J. Electrochem. Sci., 6 (2011) 3375-3384.
[4] PichananTeesetsopon, S. Kumar and JoydeepDutta, Int. J. Electrochem. Sci., 7 (2012) 4988-4999.
[5] Zhen Wei, Yu Yao, Tao Huang and Aishui Yu, Int. J. Electrochem. Sci., 6 (2011) 1871 – 1879.
[6] A.Kumar,R. Jose,K. Fujihara, J.Wang, S.Ramakrishna,Chem.Mater.19(2007) 6536-6542.
[7] W. Nuansing, S. Ninmuang,W. Jarernboon, S. Maensiri, S. Seraphin, Mater. Sci. Eng. B. 131 (2006) 147-155.

Preparation and Investigation of TiO$_2$/ZnO Composite Nanofibers for Photocatalytic Application

[1]M. V. Someshwararao, [2]R. S. Dubey*, and [3]P. S. V Subbarao

[1]Department of Physics, SRKR Engineering College, Bhimavaram, (A.P.), India

[2]Department of Nanotechnology, Swarnandhra College of Engineering and Technology, Seetharamapuram, Narsapur, (A.P.), India

[3]School Department of Physics, Andhra University, Visakhapatnam (A.P.), India

Abstract

We present the preparation and investigation of metal oxides TiO$_2$ (T), ZnO (Z) and TiO$_2$/ZnO (TZ) composite nanofibers. For the fabrication of TZ composite nanofibers, different proportions of TiO$_2$/ZnO solutions (1:1, 1:2, 2:1 and 1:3) were preferred. X-ray diffraction (XRD) patterns of the T-sample exhibited the mixed anatase and rutile phases while the wurtzite phase is examined for the Z-sample. The field-emission scanning electron microscopy (FESEM) study evidenced the preparation of continuous and randomly oriented nanofibers. Transmission electron microscopy (TEM) investigation endorsed the cylindrical morphology of the TZ13-sample with c.a. diameter 230 nm. The lattice d-spacing is estimated to be 0.298 nm which corresponds to the plane (100) of the hexagonal wurtzite ZnO while selected area electron diffraction (SAED) analysis endorsed the polycrystalline nature of the TZ13 composite nanofibers. Finally, photodegradation of the Eriochrome black T dye was performed and the catalyst TZ13 showed the significant photodegradation of the dye as compared to TZ11, TZ12 and TZ21. This boosted photocatalytic activity is ascribed to the synergetic effect of the TiO$_2$/ZnO composite nanofibers with their distinct morphology.

Keywords: TiO$_2$/ZnO Nanofibers; Electrospinning Process, X-ray diffraction, Surface Morphology.

I. Introduction

Titanium oxide (TiO$_2$) and zinc oxide (ZnO) are the equally demanded materials in several applications such as batteries, sensors, photocatalytic/water splitting, dye-sensitized solar cells etc. [1-2]. TiO$_2$ is abundant in nature, non-toxic and easy to handle. The anatase and rutile polymorphs of TiO$_2$ are the well-recognized ones due to their better stability as compared to its other forms. The anatase and rutile TiO$_2$ phases have 3.2 and 3.0 eV optical band gaps respectively. On the other hand, the ZnO material possesses better electrical properties and investigated as the alternate choice in place of the TiO$_2$ due to its almost similar band gap (3.37 eV) with its wurtzite hexagonal phase structure. Therefore, both TiO$_2$ and ZnO have been employed as the photocatalysts [3,4]. Further, for the improved photodegradation, the TiO$_2$/ZnO composite semiconductor materials are found to be promising which boosts the process of electron-hole pair separation under light irradiation. Therefore, TiO$_2$/ZnO hybrid nanomaterials in the form of particles, films, fibers etc. have been well-recognized for the photodegradation study [5,6].

For the preparation of nanofibers based on mono, composite or core-shell materials, an electrospinning technique is a simple, inexpensive and well-recognized one. Chun et al. reported the photodegradation study of dye using electrospun TiO$_2$/ZnO nanofibers as the catalyst by tuning the anatase-rutile ratio via calcination process at various temperatures. The nanofibers calcined at 650 °C having ratio anatase (48 %) to rutile (52 %) demonstrated the better photocatalytic efficiency with the rhodamine B dye [7]. Yar et al. presented the photodegradation study using catalysts based on electrospun polyacrylonitrile nanofibers decorated with the TiO$_2$, ZnO and TiO$_2$/ZnO composite nanoparticles. The degradation of malachite green dye was performed however; the hybrid nanofibers of TiO$_2$/ZnO/PAN evidenced the photocatalytic activity two times greater the bare PAN nanofibers [8]. Baek et al. fabricated and investigated the various properties of the electrospun ZnO nanofibers. The choice of calcination temperature showed a significant impact on the ZnO structure and endorsed the improved morphology at the higher temperature. In addition, the choice of higher temperature calcination led to the increased diameter of the ZnO nanofibers [9]. Li et al. reported the improved photocatalytic activity of the electrospun TiO$_2$/ZnO composite nanofibers. Further, the recycled experiment of the photodegradation showed better photodegradation efficiency and the stability of the catalyst [10]. Araujo et al. demonstrated the photodegradation study of the rhodamine B dye using catalyst based on TiO$_2$/ZnO hierarchical heteronanostructures of nanorods prepared on electrospun nanofibers. The morphological study revealed the three-dimensional arrangement of ZnO nanorods of hexagonal wurtzite on the TiO$_2$ nanoporous structure. These prepared nanostructures demonstrated as the efficient photocatalyst which showed about 90 % photodegradation of the dye within the 70 min [11]. Lotus et al. reported the fabrication of the TiO$_2$/ZnO hybrid nanofibers by electrospinning technique with their diameter in the range from 50-150 nm after the calcination. UV-vis measurement study showed two energy band gaps of TiO$_2$ at 3.0 and 3.5 eV while X-ray diffraction study exhibited the anatase and rutile phases of the TiO$_2$ and the wurtzite phase of the ZnO. This study concluded the electrospinning process as the easy fabrication technique for the TiO$_2$/ZnO hybrid nanofibers with the controllability of various properties such as structural, optical, morphological, thermal and chemical compositional [12]. Pei et al. explored the preparation of the TiO_2/ZnO nanofibers to study the photocatalytic activity. By varying the zinc acetate concentration, various composite fibers were fabricated and characterized. An optimal quantity of the catalyst was determined to increase the photocatalytic response of the composite fibers; as a result, enhanced dye degradation was noticed under visible light irradiation [13]. Liu et al. prepared the core/shell (ZnO/TiO_2) nanofibers for the photocatalytic application. X-ray diffraction study exhibited the anatase and rutile phases of the TiO_2 while hexagonal wurtzite phase was noticed corresponding for the ZnO. Comparatively, the core/shell (ZnO/TiO_2) nanofibers showed the red-shift regarded the less activation energy requirement in contrast to individual ZnO and TiO_2 nanofibers. As a result, the ZnO/TiO_2 core-shell nanofibers evidenced the enhanced photocatalytic activity along with their recyclability [14]. To study the photocatalytic activity, Li et al. presented a distinct experimental process to prepare the heterojunction ZnO/TiO_2 composite fibers. The approach was zinc plating on the electrospun TiO_2 nanofibers and then the heat treatment. As a result of these processes, the photocatalytic activity of the ZnO/TiO_2 nanofibers was found to be reasonably higher than the pure-TiO_2 nanofibers [15]. In another work, Hwang et al. employed the electrospun ZnO/TiO_2 hybrid nanofibers and studied the antimicrobial activity. A promising antimicrobial activity was investigated in the presence of gram-negative Escherichia coli and gram-positive Staphylococcus aureus under ultra-violet (UV) irradiation and in the dark as well [16]. Chen et al. studied the various properties of the ZnO/TiO_2 heterogeneous nanofibers fabricated by electrospinning process and later photocatalytic behavior was demonstrated. These nanofibers exhibited the significant degradation of the rhodamine B dye which was attributed to the reduced photo-induced charge carriers rate, improved usage of UV light and the large contact area of the catalyst [17]. Kanjwal et al. reported the photocatalytic activity of the hydrothermally treated electrospun ZnO/TiO_2 nanofibers. The photocatalytic activity of the ZnO nanoparticles prepared by hydrothermal process, electrospun TiO_2, ZnO/TiO_2 composite nanofibers and the hydrothermally treated electrospun ZnO/TiO_2 composite nanofibers were investigated and compared. The hydrothermally treated ZnO/TiO_2 nanofibers were found to be more efficient which could degrade the methyl red and rhodamine B dyes within 90 and 105 min respectively as compared to 3 h degradation time consumed by the other three catalysts [18]. Liu et al. studied the TiO_2/ZnO composite nanofibers calcined at various temperatures. To improve the photocatalytic property, the ZnO was blended in TiO_2/ZnO composite nanofibers. As a result, the photodegradation of the rhodamine B and phenol were noticed to be 100 and 85 % respectively with the 15.76 wt % ZnO content as the optimal quantity for the enhanced photocatalytic response [19]. Park et al. explored the structural and electrical properties of the electrospun ZnO nanofibers. They reported the annealing temperature as the significant factor for the improved crystallinity of the ZnO nanofibers. The electrical conductivity of the ZnO nanofibers was observed to be inversely proportional to the calcination temperature. In addition, the ZnO nanofibers was tested for the CO gas sensing ability and found to be reliable [20].

This paper presents the electrospinning fabrication and characterization of TiO_2/ZnO composite nanofibers by varying the proportion of the TiO_2/ZnO solutions. Section 2 describes the chemicals and experimental processes for the preparation of nanofibers. The characterized results and photocatalytic investigation have been presented in section 3. Finally, section 4 summarizes the work.

II. Experimental Details

Titanium Tetraisopropoxide (TTIP, Sigma-Aldrich) and Zinc Acetate Dehydrate ($Zn(CH3COO)2·2HO$, Merck) were used as the Ti and Zn precursors respectively. Catalyst: Acetic Acid (Sdfine), polymer: Polyvinyl Pyrolidone (PVP, Mw=1,300,000, Sigma-Aldrich) and the solvent: Methanol (Fisher Scientific) were used. For the photodegradation study, the Eriochrome Black T (EBT) dye was used.

For the preparation of TiO_2 solution, 0.6 ml TTIP, 10 ml methanol and 4 ml glacial acetic acid were mixed and stirred for 30 min. Later, 0.8 g PVP was added in the above solution under vigorous stirring which was continued till the preparation of the transparent and enough viscous solution. To obtain the ZnO solution, 1.2 g zinc acetate dehydrate was dissolved in 10 ml methanol followed by adding 4 ml acetic acid under stirring. Later, 0.8 g PVP was added in the ZnO solution and stirred for 1 hr to get enough viscous solution. After preparing both the solutions, composite solutions of TiO_2 and ZnO were prepared in the proportional ratio 1:1, 1:2, 2:1 and 1:3. For all the electrospinning experiments, the parameters like dc voltage, solution flow rate and the tip-collector distance were 14 kV, 1 ml/hr and 10 cm respectively. After preparation, each sample was calcined at 600 °C for 1 hr and named as TZ11, TZ12, TZ21 and TZ13 corresponding to the 1:1, 1:2, 2:1 and 1:3 proportional ratio of the TiO_2:ZnO solutions. The calcined nanofibers were characterized to investigate the phase and crystallinity using X-ray Diffraction (XRD, Bruker AXS D8 Advance, Germany), the Uv-vis absorbance using UV-Visible Spectrophotometer (UV 1800, Shimadzu, Japan), the surface morphology using field-emission scanning electron microscopy (ZIESS, Germany) and Transmission Electron Microscope (Tecnai G2 20 Twin FEI, Netherlands).

III. Results and Discussion

X-ray diffraction (XRD) study was performed to examine the phase and crystallinity of the TiO_2 (T) and ZnO (Z) nanofibers as depicted in **figure 1(a)**. The XRD pattern of the T-sample shows the presence of two crystallographic forms that are anatase (A) and rutile (R) indicating the polycrystalline nature. The anatase peaks originated at Bragg angle $2\Theta=25.2°$, $36.9°$, $37.8°$, $48°$, $53.8°$, $55°$, $62.1°$ and $68.7°$ were assigned to the planes (101), (103), (004), (200), (105), (213) and (116) respectively. Similarly, the rutile peaks originated at $2\Theta=27.4°$, $41.2°$, $44°$ and $54.3°$ were assigned to the planes (110), (111), (210) (211) respectively. The anatase and rutile peaks were found in matching with the JCPDS File No. 21-1272 and JCPDS File No. 21-1276 respectively. The XRD pattern of the Z-sample endorses the characteristic diffraction peaks of the ZnO crystallite originated at $2\Theta=31.7°$, $34.4°$, $36.2°$, $47.5°$, $56.6°$, $62.8°$ and $67.9°$ corresponding to the planes (100), (002), (101), (102), (110), (103) and (112). The XRD pattern indicates the formation of hexagonal wurtzite crystalline phase and coincides with the JCPDS File No. 36-1451.

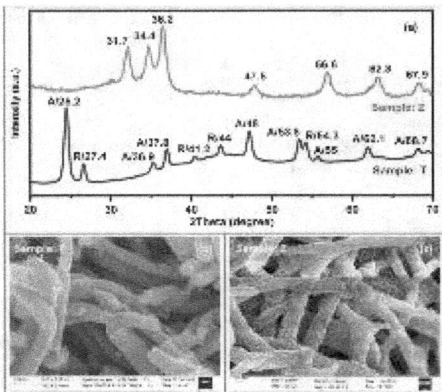

Figure 1. XRD patterns and FESEM images of TiO$_2$ (T) and ZnO (Z) nanofibers.

No other peaks were noticed in both the samples, T and Z endorsing the absence of impurities. The morphology of the TiO$_2$ and ZnO nanofibers were investigated by FESEM as shown in **figure 1(b)** and **1(c)**. The prepared nanofibers of the TiO$_2$ and ZnO were found randomly aligned. Comparatively, TiO$_2$ nanofibers possess a smoother surface than the ZnO nanofibers. Using ImageJ tool, the mean diameter of the T and Z samples were observed to be 169 and 209 nm respectively.

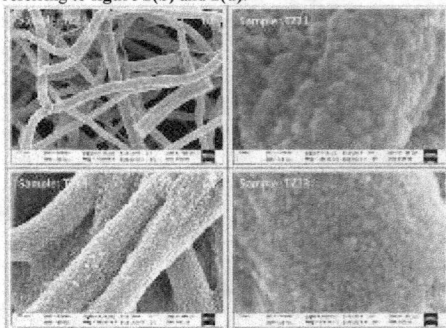

Figure 2. Surface morphology of TZ11 and TZ12 nanofibers.

Figure 2(a) and **2(b)** depicts the FESEM images of the composite TiO$_2$/ZnO nanofibers (TZ11) prepared with the proportional ratio 1:1 of the TiO$_2$:ZnO solutions. Here, we can observe the preparation of the smooth and aligned nanofibers as depicted in **figure 2(a)**. A high scale FESEM image depicted in **figure 2(b)** shows the well-aligned nanofibers with the composition of TiO$_2$ and ZnO spherical nanoparticles in the sample TZ11. Comparatively, TZ12 nanofibers were noticed to be thinner and non-uniform as depicted in **figure 2(c)**. However, **figure 2(d)** endorses the presence of TiO$_2$ and ZnO nanoparticles arranged in a random direction. The diameters of the TZ11 and TZ12 fibers were found to be in the range from 194-379 and 115-240 nm respectively. However, the average particles size was estimated to be 11 and 11.3 nm corresponding to the fiber samples TZ11 and TZ12. Here, the double proportion of the ZnO solution resulted in the rough fibers with the randomly arranged nanoparticles in the fiber form as referring to **figure 2(b)** and **2(d)**.

Figure 3. Surface morphology of TZ21 and TZ13 nanofibers.

With the increased ZnO solution i.e. proportion T:Z=1:2, we have noticed rough nanofibers as depicted in **figure 2(c)** and **2(d)**. Inversely with the sample TZ21, we can notice the preparation of the smoother nanofibers when the proportion of the TiO$_2$ solution was double than the ZnO i.e. T:Z=2:1 as shown in **figure 3(a)**. A closer look of the TZ21 fiber morphology depicted in **figure 3(b)** endorses the compact packing of the spherical nanoparticles. The diameter of the TZ21 and TZ13 fibers and nanoparticles were found to be in the range from 155-284 and 332-428 nm and 11 and 9.8 nm respectively. As depicted in **figure 3(c)**, a distinct morphology of the TZ13 nanofibers can be noticed which is based on the proportional ratio, T:Z=1:3. With a similar trend as discussed in **figure 2(c)**, we can observe the formation of rough and well-aligned nanofibers. It means roughness of the fibers is enhanced with the increased concentration of the ZnO solution. However, well packing of the TiO$_2$ and ZnO nanoparticles with somewhat porosity is observed in the FESEM image as shown in **figure 3(d)**.

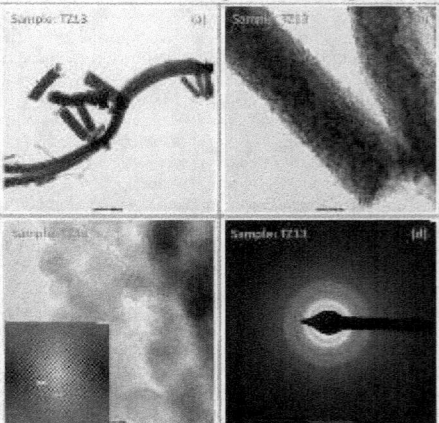

Figure 4. TEM micrographs of TZ13 nanofibers fig.(a) and fig.(b), HR-TEM fig.(c) and SAED pattern fig.(d).

With the interest of distinct morphology observed for the TZ13 sample, we have performed the transmission electron microscopy (TEM) investigation. **Figure 4(a)** shows the cylindrical morphology of the TZ13 nanofibers with somewhat rough surface, as this was noticed in **figure 3(c)**. As compared to TZ11, TZ12 and TZ21 composite nanofibers, the diameter of the TZ13 sample is found to be increased in the range 184-464 nm with the increased proportion of the ZnO solution as discussed in **figure 2** and **3**. TEM result coincides with the FESEM observation and the reported work [22]. As depicted in **figure 4(b)**, the diameter of the nanofiber is found to be c.a. 230 nm which is made-up of grain-like nanocrystals of the TiO_2 and ZnO. **Figure 4(c)** depicts the high-resolution TEM (HR-TEM) image of the TZ13 nanofibers while the d-space imaging is shown in the inset of **figure 4(c)**. The lattice d-spacing is found to be 0.298 nm which corresponds to the (100) plane of the hexagonal wurtzite ZnO as discussed later. The selected area electron diffraction (SAED) image shown in **figure 4(d)** endorses the polycrystalline nature of composite TZ13 nanofibers.

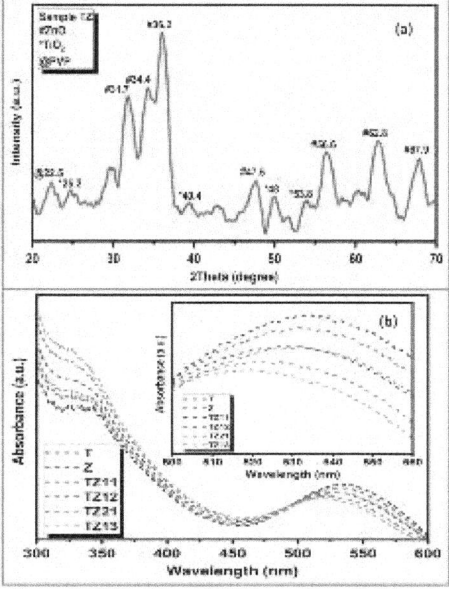

Figure 5. XRD pattern of TZ13 nanofibers fig.(a) and Uv-vis absorbance of variant composite nanofibers fig.(b).

To verify the crystallinity of the composite TZ13 nanofibers, XRD study was performed and the obtained pattern shows the good crystallinity as shown in **figure 5(a)**. The anatase TiO_2 peaks can be observed at 2Θ=25.2°, 40.4°, 48° and 53.8° while wurtzite ZnO peaks at 2Θ=31.7°, 34.4°, 36.2°, 47.5°, 56.6°, 62.8° and 67.9° are also exhibited. A peak at 2Θ=22.5° can also be observed which may be associated with the polymer residual [21]. We can recall that the XRD pattern of the T-sample endorsed the mixed crystalline phases of the anatase and rutile as shown in **figure 1(a)**. However, in the XRD pattern of the TZ13-sample only anatase phase of the TiO_2 and ZnO peaks are evidenced. More ZnO peaks in the XRD pattern is regarded the increased ZnO concentration as compared to TiO_2. Depending on the ZnO concentration, the dominant peaks may vary as reported in the literature [22]. Therefore, the d-spacing obtained by HR-TEM investigation represents the ZnO plane (100) of the hexagonal wurtzite polycrystalline phase.

Under UV illumination, the semiconductor metal-oxides like TiO_2, ZnO etc. produce oxidized hydroxyl and oxyradicals due to the generation of the electron-hole pairs and decomposes the organic materials into less dangerous contents. Using TZ11, TZ12, TZ21 and TZ13 composite fibers as the catalysts, we have studied the photodegradation of the Eriochrome black T (EBT) dye in aqueous under UV irradiation. **Figure 5(b)** depicts the UV-vis absorbance recorded in a regular interval upto 120 min of UV irradiation. A significant photodegradation of the EBT dye can be observed when the proportion of the ZnO is increased. Comparatively, TZ13-sample exhibits the best photodegradation among the TZ11, TZ12 and TZ21 samples. The anatase TiO_2 absorbs efficiently UV light and moreover, rutile phase does not exist in our case as depicted in **figure 5(a)**. Furthermore, the charge carriers recombination gets minimal due to the presence of the ZnO [22]. The improvement in the photocatalytic activity is ascribed to the synergetic effect of the TiO_2/ZnO composite nanofibers with their distinct morphology. The nanofibers possess a higher surface area than the nanoparticles which increases the active sites. As a result, this boosts the adsorption of the dye and therefore, the enhanced photocatalytic activity was exhibited [17].

IV. Conclusions

We have investigated the structural and morphological properties of the various metal oxides (T, Z, and TZ) nanofibers. The XRD patterns of the pure-TiO_2 showed the mixed anatase and rutile phases while the hexagonal wurtzite phase was noticed of the ZnO. Remarkably, XRD pattern of the TZ13 composite nanofibers exhibited the anatase peaks corresponding to the TiO_2 along with the ZnO-wurtzite peaks. The XRD pattern of the TZ13 composite nanofibers is dominated with the major peaks of the ZnO which could be regarded the increased proportion of the ZnO solution as compared to TiO_2. Morphology studied by FESEM evidenced the smooth, long and randomly aligned nanofibers for all the samples T, Z, TZ11, TZ12, TZ21 and TZ13. However, a distinct morphology of the TZ13 nanofibers is attributed to the increased proportion of the ZnO solution (T:Z=1:3). Though TZ13 nanofibers are found to be well-aligned with its c.a. diameter 230 nm but noticed to be rougher. TEM investigation endorsed the well-aligned nanofibers with the lattice d-spacing around 0.298 nm which is regarded the (100) plane of the wurtzite hexagonal phase of the ZnO. The selected area electron diffraction (SAED) analysis showed the polycrystalline nature of the TZ13 nanofibers and coincided with the XRD result. Furthermore, the photodegradation of the EBT dye using TZ11, TZ12, TZ21 and TZ13 catalysts were performed whereas TZ13-sample endorsed the best photodegradation of the EBT dye.

Acknowledgements

The authors are thankful to Prof. M. A. More, Savitribai Phule Pune University, (M.S.), India for the provided accessibility of the characterization facilities.

Competing Interests

The authors declare that they have no competing interests.

References

1. Agnieszka Kołodziejczak-Radzimska and Teofil Jesionowski, Zinc Oxide-From Synthesis to Application: A Review, Materials (Basel)., Vol. 7(4), 2833–2881 (2014).
2. Adawiyah J.Haider, Zainab N.Jameel, Imad H.M.Al-Hussaini, Review on: Titanium Dioxide Applications, Energy Procedia Vol. 157, 17-29 (2019).
3. Xueyan Li, Desong Wang, Guoxiang Cheng, Qingzhi Luob Jing An, Yanhong, Wang, Preparation of polyaniline-modified TiO2nanoparticles and their photocatalytic activity under visible light illumination, Applied Catalysis B: Environmental,Vol. 81, Issues 3–4, 267-273 (2008).
4. Navin Jain, Aprit Bhargava and Jitendra Panwar, Enhanced photocatalytic degradation of methylene blue using biologically synthesized "protein-capped" ZnO nanoparticles, Chemical Engineering Journal, Vol. 243, 549-555 (2014).
5. Jintao Tian, Lijuan Chen, Yansheng Yin, Xin Wang, Jinhui Dai, Zhibin Zhu, Xiaoyun Liu and Pingwei Wu, Photocatalyst of TiO2/ZnO nano composite film: Preparation, characterization, and photodegradation activity of methyl orange, Surface & Coatings Technology 204 (2009) 205–214.
6. D.L.Liao, C.A.Badour and B.Q.Liao, Preparation of nanosized TiO2/ZnO composite catalyst and its photocatalytic activity for degradation of methyl orange, Journal of Photochemistry and Photobiology A: Chemistry, 194/1, 11-19 (2008).
7. Carina Chun Pei andWallace Woon-Fong Leung, Enhanced photocatalytic activity of electrospun TiO2/ZnO nanofibers with optimal anatase/rutile ratio, Catalysis Communications 37 (2013) 100–104.
8. Adem Yar, Bircan Haspulat, Tugay U¨ stu¨n, Volkan Eskizeybek, Ahmet Avcı, Handan Kamı‚s and Slimane Achour, Electrospun TiO2/ZnO/PAN hybrid nanofiber membranes with efficient photocatalytic activity, RSC Adv., 7, 29806–29814 (2017,).
9. Jeong-Ha Baek, Juyun Park, Jisoo Kang, Don Kim, Sung-Wi Koh, and Yong-Cheol Kang, Fabrication and Thermal Oxidation of ZnO Nanofibers Prepared via Electrospinning Technique, Bull. Korean Chem. Soc., Vol. 33, No. 8 (2012) 2694-.
10. Jian Li, Long Yan, Yufei Wang, Yuhong Kang, Chao Wang and Shaobo Yang, Fabrication of TiO2/ZnO composite nanofibers with enhanced photocatalytic activity, J Mater Sci: Mater Electron (2016).
11. Evando S. Araújo, Bruna P. da Costa, Raquel A.P. Oliveira, Juliano Libardi, Pedro M. Faia and Helinando P. de Oliveira, TiO2/ZnO hierarchical heteronanostructures: Synthesis, characterization and application as photocatalysts, Journal of Environmental Chemical Engineering 4 (2016) 2820–2829.
12. A.F. Lotus , S.N. Tacastacas, M.J. Pinti, L.A. Britton, N. Stojilovic , R.D. Ramsier and G.G. Chase, Fabrication and characterization of TiO2–ZnO composite nanofibers, Physica E 43 (2011) 857–861
13. Carina Chun Pei and Wallace Woon-Fong Leung, Photocatalytic degradation of Rhodamine B by TiO2/ZnO nanofibers under visible-light irradiation, Separation and Purification Technology 114 (2013) 108–116.
14. Xian Liu, Yan-yu Hu, Ri-Yao Chen, Zhen Chen, and Hong-Chun Han, Coaxial Nanofibers of ZnO-TiO2 Heterojunction With High Photocatalytic Activity by Electrospinning Technique, Synthesis and Reactivity in Inorganic, Metal-Organic, and Nano-Metal Chemistry, 44:449–453, 2014
15. Delong Li, Xudong Jiang, Yupeng Zhang, and Bin Zhang, A novel route to ZnO/TiO2 heterojunction composite fibers, J. Mater. Res., Vol. 28, No. 3, Feb 14, 2013.
16. Sun Hye Hwang, Jooyoung Song, Yujung Jung, O. Young Kweon, Hee Song and Jyongsik JangElectrospun ZnO/TiO2 composite nanofibers as a bactericidal agentw, Chem. Commun., 2011, 47, 9164–9166.
17. Jia Dong Chen, Wei Sha Liao, Ying Jiang, Dan Ni Yu, Mei Ling Zou, Han Zhu, Ming Zhang and Ming Liang Du, Facile Fabrication of ZnO/TiO2 Heterogeneous Nanofibres and Their Photocatalytic Behaviour and Mechanism towards Rhodamine B, Nanomater Nanotechnol, 2016, 6:9, doi: 10.5772/62291.
18. Muzafar A. Kanjwal, Nasser A. M. Barakat, Faheem A. Sheikh, Soo Jin Park and Hak Yong Kim, Photocatalytic Activity of ZnO-TiO2 Hierarchical Nanostructure Prepared by Combined Electrospinning and Hydrothermal Techniques Macromolecular Research, Vol. 18, No. 3, pp 233-240 (2010).
19. Ruilai Liu, Huiyan Ye , Xiaopeng Xiong and Haiqing Liu, Fabrication of TiO2/ZnO composite nanofibers by electrospinning and their photocatalytic property, Materials Chemistry and Physics 121 (2010) 432-439.
20. Jin-Ah Park, Jaehyun Moon, Su-Jae Lee, Sang-Chul Lim and Taehyoung Zyung, Fabrication and characterization of ZnO nanofibers by electrospinning, Current Applied Physics 9 (2009) S210–S212.
21. M. H. Abou_Taleb, Thermal and Spectroscopic Studies of Poly(N-vinyl pyrrolidone)/Poly(vinyl alcohol) Blend Films. Journal of Applied Polymer Science, Vol. 114, 1202–1207 (2009).
22. Jia Dong Chen, Wei Sha Liao, Ying Jiang, Dan Ni Yu, Mei Ling Zou, Ming Zhang, Ming Liang Du and Han Zhu, Facile Fabrication of ZnO/TiO2 Heterogeneous Nanofibres and Their Photocatalytic Behaviour and Mechanism towards Rhodamine B, Nanomater Nanotechnol, Vol., 6/9 , pp8, doi: 10.5772/62291, (2016).

Fabrication and characterisation of electrospun barium titanate and polyvinly pyridine composite nanofibers

Ravi Bala Krishna [a,*], M.V. Someswararao [b], P.S.V. Subbarao [c], R.S. Dubey [d], B.S. Diwakar [e], K.S. Jaideep [a]

[a] Department of Mechanical Engineering, S.R.K.R. Engineering College, China-Amiram 534204, India
[b] Department of Engineering Physics, S.R.K.R. Engineering College, China-Amiram 534204, India
[c] Department of Physics, Andhra University, Visakhapatnam 530003, India
[d] Department of Nanotechnology, Swarnandra College of Engineering & Technology, Seetarampuram 534280, India
[e] Department of Engineering Chemistry, S.R.K.R. Engineering College, China-Amiram 534204, India

ARTICLE INFO

Article history:
Received 2 May 2019
Received in revised form 19 June 2019
Accepted 21 June 2019
Available online 16 July 2019

Keywords:
Barium titanate (BaTiO$_3$)
Poly vinyl pyrolidene (pvp)
Electrospinning
Scanning Electron Microscope (SEM)
X-ray diffraction (XRD)

ABSTRACT

BaTiO$_3$-PVP composite nanofibers were successfully produced by sol-gel and Electrospinning method. These fibers were calcinated at 500 °C for 2 h and characterized by XRD, SEM, FTIR, and TG-DTA. XRD measurements confirm that the existence of the pure anatase phase of (BaTiO$_3$-PVP) fibers. FTIR studies reveals that the formation of metal oxide bond at 570–600 cm^{-1}. TG-DTA analysis of BaTiO$_3$-PVP composite nanofibers indicate that most of the acetate and organic groups were removed approximately at 700 °C. SEM studies shows the fibers formed with diameter in the range of 200 nm. Finally electrospun BaTiO$_3$-PVP nanofibers diameter is decreased with increasing the applied voltage.

© 2019 Elsevier Ltd. All rights reserved.
Selection and peer-review under responsibility of the scientific committee of the 2nd International Conference on Applied Sciences and Technology (ICAST-2019): Materials Science.

Introduction to Barium titanate (BaTiO$_3$)

Barium Titanate (BaTiO$_3$) is one of the very important lead free ferroelectric material with large number of applications such as transducers, dielectric capacitors, in non-volatile ferroelectric random access memories, sensors and actuators, in solid oxide fuel cells, etc. [1,2]. If these ferroelectric are prepared in nanosize we expected to improve their properties. Nanostructured BaTiO$_3$ fibers wires and tubes are very attractive because of their high surface area to volume ratio. BaTiO$_3$ ferroelectrics are polar materials that exhibit net spontaneous polarization and hysteresis behavior in external applied electricfield. BaTiO$_3$ with perovskite structure is widely used in various electric applications such as Multilayer Ceramic Capacitors (MLCCS), and ferroelectric random access memories, etc. [3,4]. Recently BaTiO$_3$with pvp composite nanofibers were fabricated by using Electrospinning method with one dopants also because Electrospinning is a very simple and straight forward technique for the preparation of polymer and oxide nanofibers. Initially this technique was used for the preparation of polymer nanofibers [5–8]. From last few years this technique has been used for the preparation of metal oxide/ceramic nanofibers such as Titania, Aluminia, Silica, Zircon, Nickel oxide, Tin oxide, Lead zirconate titanate and other oxide materials [9–11].

Electrospinning apparatus consisting of a syringe pump, metal needle, high voltage power supply and a grounded collector etc. This Electrospinning process involves the application of a strong electrostatic field to a polymer solution filled in a syringe with metallic needle. The positive terminal of a high D. C. source is connected to the counter electrode in the form of a metal plate or aluminium foil placed at a fixed distance.

Under the influence of the electrostatic field, the solution experiences repulsive force. As the voltages surpass a threshold value, electrostatic forces overcome the surface tension and a fine charged jet is ejected. The jet moves towards the counter electrode and subdivides into large numbers due to High repulsive force, finally deposits in the form of nanofibers on the counter electrode [12–14]. In our work we are focused on fabrication and characterization of BaTiO$_3$-PVP composite nano fibers with different applied voltages by keeping flow rate and jet distance at constant and studied the effect of voltage on the morphology of the nanofibers.

https://doi.org/10.1016/j.matpr.2019.06.653
2214-7853/© 2019 Elsevier Ltd. All rights reserved.
Selection and peer-review under responsibility of the scientific committee of the 2nd International Conference on Applied Sciences and Technology (ICAST-2019): Materials Science.

2. Preparation and experimental procedure

Materials used in this are Barium acetate, ethyl alcohol, glacial acetic acid, and Titanium-IV butoxide, PVP (M.wt ~ 1,300,000) chemicals are taken from Sigma Aldrich (99%). The composite precursor solution was prepared by using Barium Titanate and pvp sol–gel. In this work, two solutions are prepared named as SOLUTION-A and SOLUTION-B with different concentrations and their details are given below.

2.1. SOLUTION-A

First, 1.92 g of barium acetate (Sigma Aldrich, 99%) was added in a 25 ml breaker with 4.5 ml of acetic acid (Pharmaco-Aaper, 99.7%). The above solution was mixed thoroughly with the help of a magnetic stirrer. Three drops of deionized (DI) water was added into this solution to ensure the barium acetate be completely dissolved and then 2.55 ml of Titanium (IV) butoxide is added slowly to this solution.

2.2. SOLUTION-B

Take another beaker of 25 ml, add 5 ml of ethanol, followed by 0.4 g of Poly vinyl pyrrolidone (pvp) (Sigma Aldrich, M. Wt ~ 1,300,000) which is used to adjust the viscosity and provide the carrier for the formation of nanofibers, and this solution is stirred continuously. Then SOLUTION-A is added to SOLUTION-B and this mixture was stirred continuously until it became viscous and homogeneous mixture and which is filled in the syringe pump of the Electrospinning apparatus.

2.3. Electrospinning of PVP-BaTiO$_3$ composite nanofibers

The Electrospinning is carried out by maintaining flow rate at constant and varying the distance from tip to collector (TCD) and D.C voltage as shown in Table 1. The flow rate of the solution was maintained at 1 ml/h. The composite nano fibers of PVP-BaTiO$_3$ were prepared by injecting the solution into the Electro spinning machine.

3. Characterisations

3.1. X-Ray Diffraction (XRD) analysis

XRD measurements were carried out for the investigation of crystallinity and phase analysis of the prepared BaTiO$_3$-PVP nanofibers calcinated at 500 °C which are shown in Fig. 1 as (AB1, AB2, AB3). However strong intensity peak can be observed at around 2θ = 24°, 27°. This peak confirms that the existence of the pure anatase phase in all samples which shows better crystallinity of the nanofibers and further the peaks at 2θ = 33°, 42°, 55° which represents perovskite phase of the samples.

From the XRD results it is concluded that as the voltage increases the diameter of the fiber is expected to become smaller due to which the main diffraction peak at 2θ = 27° is shifting upwards as shown in the above Fig. 1

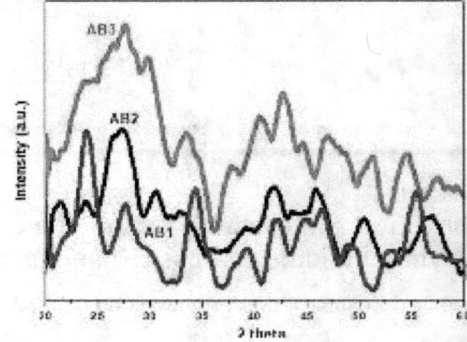

Fig. 1. XRD of all the samples.

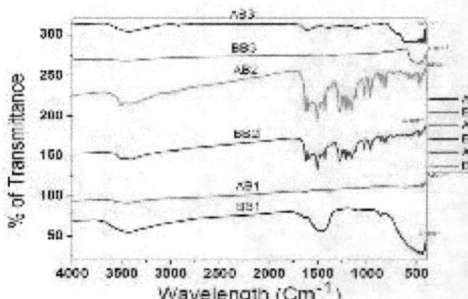

Fig. 2. FTIR spectroscopy of all the samples.

3.2. Fourier Transform Infrared Spectroscopy (FTIR)

The FTIR studies of BaTiO$_3$-PVP nanofibers as shown in below Fig. 2 represents main bonds at around 3429, 2935, 1425 a 570 cm^{-1} corresponding to O—H, C—H, O—H and Ti—O stretch Vibrations respectively.

A broad band at 570 cm^{-1}, Ti—O vibration, became sharper a narrower after calcinations at 500 °C due to the formation of me oxide bond. The presence of a peak at 570 cm^{-1} corresponding Ti—O vibration of nanofibers confirm the formation of BaTi Ti—O bonding stretching vibrations before calcination named (BB1, BB2, BB3) is occurred at around 571cm^{-1} and after calcin tion (AB1, AB2, AB3) is occurred at around 540 cm^{-1}. As the volta increases the Ti—O bonding stretching vibrations are obtained increasing wave numbers from 570 cm^{-1} to 600 cm^{-1}. The m important peak for the barium titanate and corresponding wavelength range between 530 cm^{-1} to 680 cm^{-1}. This peak due to vibrations of the bond between titanium and oxygen (Ti-bond). The bond between the Ba—Ti—O was obtained at 866 cm

Table 1

Sample no.	Run Time	Distance from tip of needle to collector (cm)	Flow rate (ml/h)	Voltage (KV)	Speed (rpm)	Humidity (%)	Room temp (°C)
1	2 h	8.5	1.0	7.5	1200	39	28
2	2 h	8.5	1.0	9.5	1200	40	28
3	2 h	8.5	1.0	11.5	1200	39	28

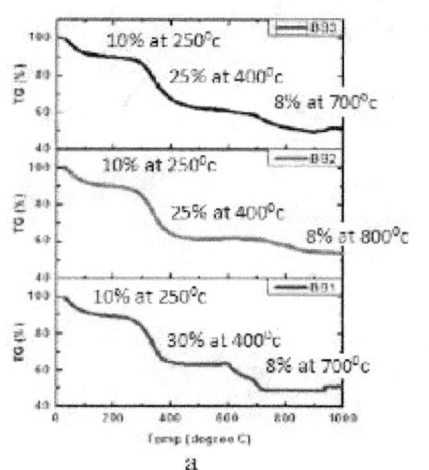 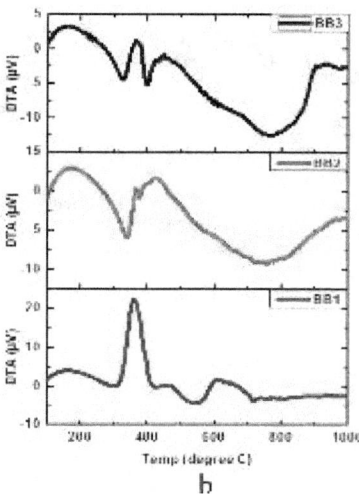

Fig. 3. (a and b) Thermogravimetric & Differential Thermal Analysis (TGDTA) of three samples.

Sample name	Labelling (before calcination)	Weight loss at different temperatures			Total weight loss
BB1	10% at 250 °C	30% from 250 to 400 °C	8% from 600 to 700 °C		48%
BB2	10% at 250 °C	25% from 250 to 400 °C	8% above 400 °C		43%
BB3	10% at 300 °C	20% from 300 to 400 °C	10% above 700 °C		40%

ore calcination, but after calcination it will be increased to cm^{-1} in the above samples. The confirmation of Ba—Ti—O ding is at around 859 cm^{-1}, C—O bond is obtained at 53 cm^{-1} and O—H bond is obtained at 3430 cm^{-1}. Finally as voltage increased the formation of metal-oxide bond can be erved as slightly increased wave number from 570 cm^{-1} to) cm^{-1}.

Thermo Gravimetric & Differential Thermal Analysis (TG-DTA)

As shown in Fig. 3a and b, below indicates that TG-DTA analysis electro spun barium titanate nanofibers. Their weight loss with pect to temperature is observed in TGA graphs from Fig. 3a. The ual weight loss as per TGA graph is 48–40% from first to third ple (BB1, BB2, and BB3) which is decreased as voltage reased (see Table 2).

It is observed from the first sample shown in the above Fig. 3a. As BB1 that the minor weight loss below 200 °C is due to the removal of traces of solvents/moisture. The weight loss between 250 °C and 400 °C is due to the decomposition of PVP and acetate molecules [15]. Similarly, the weight loss during (600–700 °C) is considered due to the decomposition of organic groups from organo-metallic precursor and its intermediate phases (titanium tetra but oxide). The weight loss was completed at 700 °C and was about 48% of the total weight.

For the second sample as BB2 the weight loss with 10% is observed slowly up to 250 °C and then from (250 to 400 °C) weight loss is around 25%. But above 400 °C the weight loss is about 8%. The total weight loss is about 43%.

- For the third sample as BB3 Weight loss with 10% is observed up to 300 °C, from 300 to 400 °C the weight loss is around 20%, and above 700 °C the weight loss is around 10%, and total weight loss about 40%. As the voltage increased, from 7.5KV to11.5KV the weight loss of the samples decreased from 48% to 40%. This shows $BaTiO_3$ nanofibers with stands high temperatures.
- It is observed from above Fig. 3b DTA analysis reveals that endothermic peaks observed at 300 °C because of evaporation of water and trapped solvent. Exothermic peaks were observed at 359 °C, 468 °C and 624 °C. Weight loss occurred at 359 °C and 468 °C probably due to the decomposition of Barium Acetate [16]. Another exothermic peak at 624 °C may correspond to the decomposition of main chain of PVP. So the TG/DTA results revealed that most of the organic groups were a vanished approximately at 700 °C. As the voltage increased from first to third sample the endothermic peaks should be changed from 300 °C and 325 °C, but exothermic peaks 359 °C, 468 °C and 624 °C are slightly changed.

3.4. SEM measurements

From the below Fig. 4 SEM measurements confirm that, the formation of $BaTiO_3$ nanofiber with different diameters before and after calcination of the samples at 500 °C. It is observed that before calcination the diameter of the fiber is larger and, after calcination, the diameter of the fiber is decreased and the minimum diameter of the fiber is 200 nm. As the voltage increased from 7.5 kV to 11.5 kV the diameter of the fiber is decreased. This is because of the evaporation of pvp, acetate and organic groups from the $BaTiO_3$ nanofibers.

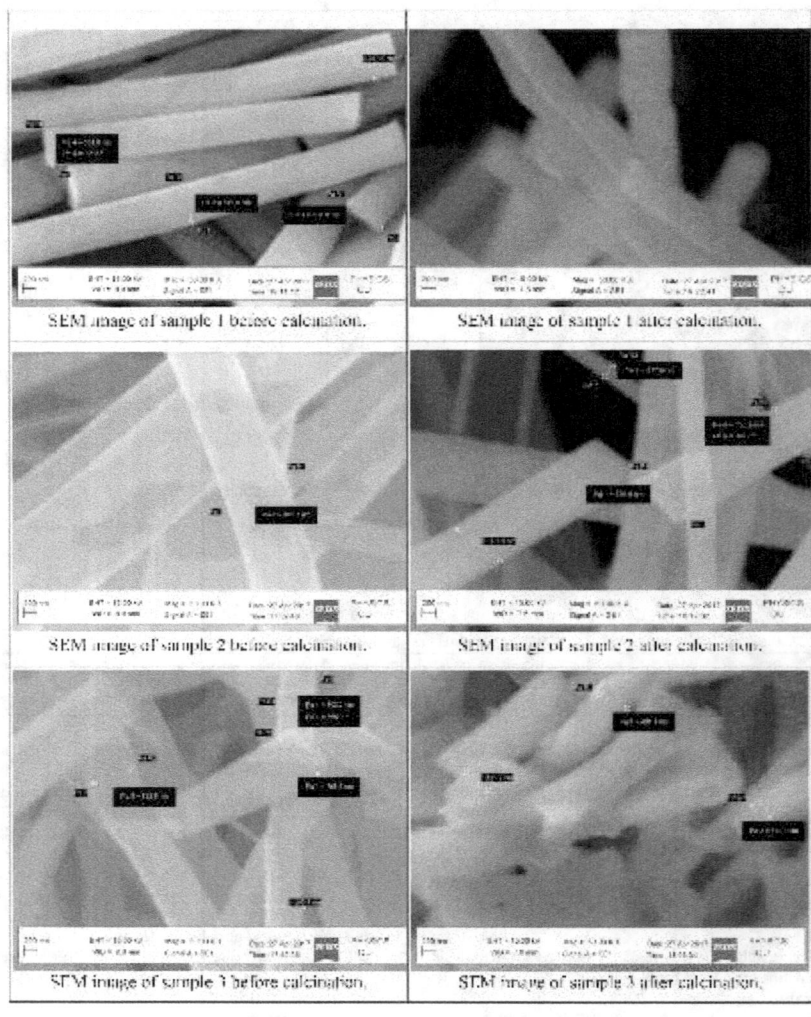

Fig. 4. SEM images of all samples before and after calcination.

4. Conclusions

Electro spun PVP-BaTiO$_3$ composite nanofibers were produced with diameter 200 nm. XRD measurements confirm that the existence of the pure anatase phase of (BaTiO$_3$-PVP) nanofibers. And from SEM measurements the diameter of the BaTiO$_3$ nanofibers decreased with the increasing the applied voltage. TG/DTA analysis of BaTiO$_3$-PVP composite nanofibers indicate that most of the organic groups were vanished approximately at 700 °C. FTIR study confirms that the formation of metal oxide bond at 570–600 cm^{-1}. Finally we have produced Electrospun BaTiO$_3$/PVP nanofibers with diameter range of 200 nm, which were utilized in the optical fiber and energy applications.

Acknowledgments

This paper is partially supported by CCPD Funds, S.R.K Engineering College.

References

[1] S.O. Brien, L. Brus, C.B. Murray, and Synthesis of monodisperse nanoparticle barium titanate: towards a generalized strategy of oxide nanopart synthesis, J. Am. Chem. Soc. 23 (2001) 12085–12086.
[2] D.H. Yoon, B.I. Lee, BaTiO$_3$ properties and powder characteristics for cera capacitors, J. Ceram. Process Res, 3 (2002) 41–47.
[3] J. Yuh, J.C. Nino, W.M. Sigmund, Synthesis of barium titanate (BaT nanofibers via electrospinning, Mater. Lett. 59 (2005) 3645.

[1] J. Yuh, L. Perez, W.M. Sigmund, J.C. Nino, Electrospinning of complex oxide nanofibers, Physica E 37 (2007) 254.
[2] D. Li, J.T. McCann, Y.N. Xia, Electrospinning: a simple and versatile technique for producing ceramic nanofibers and nanotubes, J. Am. Ceram. Soc. 89 (2006) 1861–1869.
[3] S. Sundarrajan, A.R. Chandrasekaran, S. Ramakrishna, An update on nanomaterials-based textiles for protection and decontamination, J. Am. Ceram. Soc. 93 (2010) 3955–3975.
[4] P.K. Panda, Ceramic nanofibers by electrospinning technique: a review, Trans. Ind. Ceram. Soc. 66 (2008) 65–76.
[5] Z. Huang, Y. Zhang, M. Kotaki, S. Ramakrishna, A review on polymer nanofibers by electrospinning and their applications in nanocomposites, Comp. Sci. Technol. 63 (2003) 2223–2253.
[6] W. Sigmund, J. Yuh, H. Park, V. Maneeratana, G. Pyrgiotakis, A. Daga, J. Taylor, J. C. Nino, Processing and structure relationships in electrospinning of ceramic fiber systems, J. Am. Ceram. Soc. 89 (2006) 395–407.
[10] P.K. Panda, S. Ramakrishna, Electrospinning of alumina nanofibers using different precursors, J. Mater. Sci. 42 (2007) 2189–2193.
[11] H. Wu, W. Pan, Preparation of zinc oxide nanofibers by electrospinning, J. Am. Ceram. Soc. 89 (2006) 699–701.
[12] H. Li, W. Zhang, B. Li, W. Pan, High Tc in electrospun $BaTiO_3$ nanofibers, J. Am. Ceram. Soc. 92 (2009) 2162–2164.
[13] J.T. Mc Cann, J.I.L. Chen, D. Li, Z.G. Ye, Y.A. Xia, Electrospinning of polycrystalline barium titanate nanofibers with controllable morphology and alignment, Chem. Phys. Lett. 424 (2006) 162–166.
[14] S. Kumar, G.L. Messing, W.B. White, Metal organic resin derived barium titanate. I. Formation of barium titanate oxycarbonate intermediate, J. Am. Ceram. Soc. 76 (1993) 617–624.
[15] J.Y. Junhan, L. Perez, W.M. Sigmund, C.N. Juan, Sol–gel based synthesis of complex oxide nanofibers, J. Sol–Gel Sci. Technol. 42 (2007) 323–329.
[16] D.P. Birnie, Esterification kinetics in titanium isopropoxide–acetic acid solutions, J. Mater. Sci. 35 (2000) 367–374.

Chapter-6

Conclusions and Future Remarks

6.1 Conclusions

An electrospinning approach is a versatile and simple technique for the fabrication of nanofibers for their potential applications. The prepared samples were studied by x-ray diffraction (XRD) and endorsed the mixed phases of anatase and rutile-TiO_2. TG/DTA study of TiO_2/PVP mat demonstrated the thermal response with respect to increased temperature. By performing the FTIR study, the presence of various functional groups along with the vibration peak corresponding to the Ti-O-Ti bond is found at 660 cm^{-1}. Before the optimization of process parameters, FESEM analysis revealed the formation of randomly distributed nanofibers with their diameter in the range from 244-343 nm. An increased applied voltage and distance from tip-collector respectively produced thinner nanofibers in the range from 293 nm-175 nm and 259 nm-147 nm.

Further, the formation of thin nanofibers of diameter 214 nm-111 nm and 284 nm-102 nm are observed with the decreased flow rate of solution and PVP concentration respectively. Finally, the optimized values of process parameters of electrospinning could decrease the diameter of TiO_2 nanofibers to 74 nm. DSSC photoanode was fabricated using thinner TiO_2 nanofibers, and the photovoltaic performance was evaluated. This process parameter study is beneficial to tailor diameter of the nanofibers either thin or thick according to application.

Furthermore, the work was extended to fabricate ZnO (Z) and TiO$_2$/ZnO (TZ) composite nanofibers along with TiO2 (T). The XRD patterns of pure-TiO$_2$ (T) revealed the mixed anatase and rutile phases while the ZnO (Z) sample evidenced the hexagonal wurtzite structure. To study the morphology of the TiO$_2$/ZnO (TZ) composite nanofibers, the concentration of TiO$_2$ and ZnO solutions were varied. Significantly, the TZ13 composite nanofibers endorsed the TiO$_2$-anatase peaks along with the ZnO-wurtzite peaks as revealed by the XRD pattern. In the case of TZ13 composite nanofibers, XRD pattern is found to be dominated by the ZnO peaks as compared to TiO$_2$ which is attributed to the increased concentration of ZnO solution.

FESEM Morphology endorsed that the samples T, Z, TZ11, TZ12, TZ21, and TZ13 were smooth, long, and randomly aligned nanofibers. Nonetheless, a distinct morphology of TZ13 nanofibers is due to the increased ZnO solution (T: Z=1:3). TEM investigation evidenced the formation of well-aligned nanofibers with the lattice d-spacing about 0.298 nm, which is designated the (100) ZnO hexagonal-wurtzite plane. The study of selected area electron diffraction (SAED) revealed the polycrystalline nature of TZ13 nanofibers and found in good agreement with the XRD pattern. In addition, photodegradation was carried out for the EBT dye using catalysts TZ11, TZ12, TZ21, and TZ13, whereas the TZ13 sample endorsed the best photodegradation for the EBT dye.

6.2 Future Remarks

Based on its distinct physical and chemical properties of 1D nanostructure 'nanofibers' have their potential demand in the context of scientific, medical, and industrial applications. Nanofiber is a type of nanomaterial whose diameter can be scaled from tens to hundreds of nanometer which makes these materials unique due to its large surface-to-volume ratio. Nanofibers have the ability to produce the extremely porous mesh which could be periodically fabricated with air voids for their amazing applications. The substantial impact of nanofibers technology can indeed be ascribed by the variety of basic materials that can be used for the fabrication of nanofibers.

In other words, fundamental materials like natural polymers, synthetic polymers, carbon substances, semiconductor substances and hybrid materials can be used to realize the advanced applications. The present scenario of nanofibers demand has been extended in commercial products like fabrics, filtering, wound/cut repairing, human body organs like knee implantation and many more. Nanofibers are being recognized as promising materials dye-sensitized solar cells, energy storage, batteries, pollutant/water treatment, environmental control, medical surgery, etc.

In this context, our work is useful to understand the tuning of physical, chemical, mechanical, thermal and biological properties by considering the electrospinning parameters in order to achieve the better performance of the advanced devices based on either single „TiO$_2$' or composite „TiO$_2$/ZnO' nanofibers. This thesis is having the scope for the new comer scholar where he/she can use this work for the further application point of view.

www.ingramcontent.com/pod-product-compliance
Lightning Source LLC
Chambersburg PA
CBHW050359120526
44590CB00015B/1754